5

APPUNTI

Riccardo Barbieri
Scuola Normale Superiore and INFN
Piazza dei Cavalieri, 7
56126 Pisa, Italy

Lectures on the ElectroWeak Interactions

Riccardo Barbieri

Lectures on the ElectroWeak Interactions

EDIZIONI
DELLA
NORMALE

© 2007 Scuola Normale Superiore Pisa

ISBN 978-88-7642-311-6

To my family,
which includes as well Bruno,
Clara and Emma

Contents

Preface   **XI**

1 From the Fermi Theory of β-decay to the minimal Gauge Lagrangian   **1**
   1.1   The Fermi theory of neutron decay   1
   1.2   From the Fermi theory to the minimal gauge Lagrangian   3

2 The Lagrangian of the standard model, including neutrino masses   **7**
   2.1   The global symmetries of the minimal gauge Lagrangian   7
   2.2   Towards a realistic Lagrangian   8
   2.3   The accidental symmetries of the Standard Model   9

3 The main predictions of the standard model   **11**
   3.1   Gauge symmetry breaking and particle masses   11
   3.2   Couplings to fermions of the gauge bosons   13
   3.3   The Higgs boson   13

4 Precision tests   **15**
   4.1   Parity violation in atomic physics   15
   4.2   Leading corrections to the ρ parameter   16
   4.3   Sensitivity to the Higgs mass   18
   4.4   Vacuum polarization amplitudes in a general *universal* theory   22
   4.5   Current experimental constraints   23
   4.6   An interlude: making it without a Higgs boson   24

5 Flavour physics   **27**
   5.1   The theorems of flavour physics   27
   5.2   Individual lepton number conservation   28

	5.3	About the unitarity of the Cabibbo-Kobaiashi-Maskawa matrix	30
	5.4	Calculable flavour changing neutral current processes	31
	5.5	Summary of calculable FCNC processes	34

6 CP violation   37
    6.1 The source(s) of CP violation in the Lagrangian of the Standard Model 37
    6.2 Electric dipole moments 39
    6.3 CP violation in effective 4-fermion interactions 41

7 Basics of neutrino physics   45
    7.1 The three options for neutrino masses in the Standard Model 45
    7.2 The physical parameters 46
    7.3 Neutrino mass measurements from the β-decay spectrum 48
    7.4 Neutrino-less double-β decay 48

8 Neutrino oscillations   51
    8.1 Neutrino oscillations in vacuum 51
    8.2 Neutrino propagation in matter 53
    8.3 Current determination of neutrino masses and mixings . 55

9 The naturalness problem of the Fermi scale   57
    9.1 The Standard Model as a prototype effective theory . . . 57
    9.2 Expanding in operators of higher dimension 58
    9.3 Minimal Flavour Violation 60
    9.4 The naturalness scale of the Standard Model 61
    9.5 The *little hierarchy* problem 62

10 The main drawback of the Standard Model   65
    10.1 Gauge anomalies and charge quantization 65
    10.2 The unification way 67

A General structure of a gauge theory   69

B Real and chiral representations of the gauge group $\mathcal{G}$ 71

C Spontaneous breaking of a gauge or a global symmetry   73

D Renormalizable theories and effective theories   75

E CP invariance   77

F Weyl, Dirac and Majorana neutrinos   79

G Anomalies   81

Preface

Elementary particle physics is the quadrant of nature whose laws can be written in a few lines with absolute precision and the greatest empirical adequacy. If this is the case, as I believe it is, it must be possible and is probably useful to introduce the students and the interested readers to the entire subject in a compact way. This is the main aim of these Lectures.

The Standard Model is the reference theory for particle physics, including the fact that one often explicitly refers to *Beyond the Standard Model* physics. Although maybe practical, I have never liked the distinction between Standard Model and Beyond the Standard Model physics. These lectures are certainly mostly about the Standard Model, minimally extended to include neutrino masses. As such, I avoid discussing explicitly any proposal that goes beyond the Standard Model, none of which has received yet any clear experimental confirmation. Nevertheless most of the Lectures are given with an open eye to a possible evolution of the theory of the ElectroWeak Interactions. At least I try. It will probably be most useful to read and use the Lectures with the same spirit. Not unrelated to this is the fact that, while these Lectures are being written, the commissioning of the Large Hadron Collider in Geneva is close to completion. Is it not risky, then, to write lectures about the ElectroWeak Interactions precisely when the incoming experiments at the LHC may demand a strong revision of the underlying theory? Maybe yes, but I think it is nevertheless useful to take such a risk now, at least as a way to focus on the open questions that the LHC experiments might allow to answer.

There are two main difficulties in trying to give a concise course on theoretical particle physics. The first one is the number of different specialized chapters that compose nowadays particle physics. Since I think that one should resist to excessive specialization, the only (not negligible) sacrifice that I try to make in compiling the list of subjects is to leave out a discussion of strong interactions. Partially this is because I consider the Lagrangian of QuantumCromoDynamics more likely to be established than the ElectroWeak sector of the Standard Model, although not to be confused with the statement that there are no important open problems in the physics of the strong interactions. The second difficulty is of technical nature, as experienced by anybody lecturing on the subject. There is a good deal of field theory that the students should know to appreciate at best a course on theoretical particle physics. Probably as a consequence of this, several excellent books on field theory exist that include a description of the Standard Model only towards the end or at least in their second part. To be able to focus as concisely as possible on physics issues, I prefer to avoid any introduction on field theory. For this reason I include a number of short Appendices that summarize the field theory knowledge that is needed for a full understanding of the content of the various Lectures. Needless to say these Appendices cannot replace a course in field theory. Again for reasons of conciseness I skip several technical details, perhaps more than usual in a pedagogical booklet. As a partial remedy, I set problems in the course of the Lectures, without solving them explicitly. A student interested in becoming able to actively work in the field of particle physics should try to solve as many of them as possible.

I thank all my collaborators, in particular Guido Altarelli, Francesco Caravaglios, Paolo Ciafaloni, George Dvali, Gian Giudice, Lawrence Hall, Luciano Maiani, Michelangelo Mangano, Antonio Masiero, Yasunori Nomura, Riccardo Rattazzi, Andrea Romanino, Giovanni Ridolfi, Slava Rychkov and Alessandro Strumia for the many interactions, discussions, corrections of errors, etc. that have influenced my understanding and my view of the theory of the electroweak interactions.

Finally I apologize for not providing any bibliography of the literature, which would have to be very large to be complete.

Riccardo Barbieri

Pisa, September 2007

Chapter 1
From the Fermi Theory of β-decay to the minimal Gauge Lagrangian

1.1. The Fermi theory of neutron decay

In 1934 Fermi wrote the first effective Lagrangian to describe a weak interaction phenomenon: nuclear β-decay. Considering neutron decay, the Fermi Lagrangian can be written, with current knowledge, as

$$\mathcal{L}_F = \frac{G_F}{\sqrt{2}} \cos\theta_C (\bar{p}\gamma_\mu(1+\alpha\gamma_5)n)(\bar{e}\gamma_\mu(1-\gamma_5)\nu), \qquad (1.1)$$

where p, n, e, ν are the fields of the proton, neutron, electron and neutrino respectively. In units where $\hbar = c = 1$, G_F is a constant of dimension of $mass^{-2}$, since every fermion field, as readily seen from the free Lagrangian, has dimension of $mass^{3/2}$. The presence of the factor $\cos\theta_C$, close to unity, will be commented later on. Finally α is another dimensionless constant.

This interaction allows to calculate both the neutron width and its angular dependence. For the total width one finds

$$\Gamma = \frac{G_F^2 \Delta^5}{60\pi^3} \cos^2\theta_C (1+3\alpha^2)\Phi, \qquad (1.2)$$

where $\Delta = 1.29$ MeV is the neutron-proton mass difference and $\Phi = 0.47$ is a numerical factor that would be unity if the electron mass were neglected relative to Δ. The angular dependence of the width in the neutron rest frame is given by

$$\frac{d\Gamma}{d\Omega_e} \propto \left(1 + \frac{1-\alpha^2}{1+3\alpha^2}\mathbf{v}_e \cdot \mathbf{n}\right), \qquad (1.3)$$

where $\mathbf{v}_e$ and $\mathbf{n}$ are the 3-velocities of the electron and of the neutrino respectively. [*Problem 1.1.1: Prove equations (1.2) and (1.3) starting from equation (1.1).*]

The measurements of the lifetime and of the angular distribution give

$$\tau = \frac{1}{\Gamma} = 885.7 \pm 0.8 \text{ s}, \qquad \alpha = -1.2695 \pm 0.0029, \qquad (1.4)$$

from which one infers, using equation (1.2), $G_F^{-1/2} \approx 250$ GeV. As we shall see, this "Fermi scale" plays a fundamental role in the theory of the electroweak interactions. It is believed to be one of the two fundamental scales in particle physics, the other being the scale of the strong interactions or the scale of Quantum CromoDynamics.

Based on the analogy with Quantum ElectroDynamics, Fermi himself, among others, conjectured that the interaction in equation (1.1) could result from the exchange of a heavy charged vector boson, $W_\mu^\pm$, of mass m_W, interacting with the current

$$J_\mu^- = \bar{p}\gamma_\mu \frac{1+\alpha\gamma_5}{2} n + \bar{\nu}\gamma_\mu \frac{1-\gamma_5}{2} e, \quad J_\mu^+ = (J_\mu^-)^+ \qquad (1.5)$$

via

$$\mathcal{L}_{\text{int}} = \frac{g}{\sqrt{2}} W_\mu^+ J_\mu^- + h.c., \qquad (1.6)$$

where g is a dimensionless coupling. Since Δ is negligible with respect to m_W, the exchange of the W-boson gives indeed rise to the Fermi interaction in equation (1.1) with the identification

$$\frac{G_F}{\sqrt{2}} = \frac{g^2}{8m_W^2}. \qquad (1.7)$$

Other effects due to the exchange of the W would also be of purely leptonic and purely hadronic nature. Limiting g to a value of order unity to maintain the perturbative analogy with QED gives an upper bound to the W-mass of order 100 GeV.

We now know that the elementary interactions of the W are not with the neutron and the proton but rather with the quark fields, u and d, through the current

$$J_\mu^- = \bar{u}\gamma_\mu \frac{1-\gamma_5}{2} d + \bar{\nu}\gamma_\mu \frac{1-\gamma_5}{2} e, \qquad (1.8)$$

which only involves the left-handed spinor fields. It is this interaction, therefore, that we want to describe in an overall consistent framework.

1.2. From the Fermi theory to the minimal gauge Lagrangian

The only known way to make sense in perturbative field theory of an interacting vector boson is to promote it to be the carrier of a gauge interaction, as described by a gauge Lagrangian $\mathcal{L}_g$ (see Appendix A). For any vector boson A_μ^A there is an associated generator T^A of the gauge group $\mathcal{G}$ forming a closed algebra

$$[T_A, T_B] = if_{ABC}T_C, \quad (1.9)$$

where f_{ABC} are the structure constants of the group. If we denote with Ψ_α, $\alpha = 1, 2, ..., N_\Psi$, the collection of the fermionic fields, all conventionally taken as left-handed Weyl spinors (a right-handed spinor can be made left-handed by charge conjugation), the gauge bosons interact with them via the current (while keeping the same notation, here the T_A are the particular matrices which represent the generators as acting on the Ψ)

$$J_\mu^A = \bar{\Psi}\gamma_\mu T^A \Psi. \quad (1.10)$$

The problem therefore is to embed equation (1.8) into a structure of the type (1.10).

For this purpose we rewrite the current in (1.8) in the form

$$J_\mu^\pm = \bar{Q}_L \gamma_\mu \frac{\sigma^\pm}{2} Q_L + \bar{L}_L \gamma_\mu \frac{\sigma^\pm}{2} L_L \quad (1.11)$$

where σ_i are the usual Pauli matrices, $\sigma^\pm = 1/\sqrt{2}(\sigma_1 \pm \sigma_2)$, and we have organized the left handed fermions into *doublets*

$$Q_L = \frac{1-\gamma_5}{2}\begin{pmatrix} u \\ d \end{pmatrix}; \quad L_L = \frac{1-\gamma_5}{2}\begin{pmatrix} \nu \\ e \end{pmatrix}. \quad (1.12)$$

So far this is only a notational change. On the other hand, by comparison with equation (1.10), we can try to identify the matrices $\sigma^\pm/2$ with the generators $T^\pm$ of a suitable gauge group. For this to be possible at all, the algebra of the generators (1.9) must be closed, which leads us to consider the commutator

$$[T^+, T^-] = [\frac{\sigma^+}{2}, \frac{\sigma^-}{2}] = i\frac{\sigma_3}{2} \equiv iT_3 \quad (1.13)$$

as the third generator of an $SU(2)$ algebra. To embed the *charged* current in (1.8) into a gauge structure, we have to include also a gauge boson interacting with a *neutral* current, i.e. through the diagonal generator T_3.

In the sixties, when this theory was formulated, one such *neutral* gauge boson was of course already known: the photon, whose associated diagonal generator is the electric charge Q. However the eigenvalues of Q differ from those of T_3 and, furthermore, Q acts also on the right handed components of the charged fermions (or their charged conjugates, according to our convention), whereas the T_i only act on the left handed fermions, as dictated phenomenologically by (1.8). What then if we commute the electric charge Q with the T_i? To answer this question, it is best to rewrite the electric charge generator as $Q = T_3 + Y$, where the *hypercharge* Y is another neutral generator with eigenvalues ($T_3 = 0$ on the right handed fields)

	u_L	d_L	$(u_R)^C$	$(d_R)^C$	ν_L	e_L	$(e_R)^C$
Y	1/6	1/6	−2/3	1/3	−1/2	−1/2	1

(1.14)

Note now that the hypercharge does not distinguish the individual members of the doublets, Q_L or L_L, hence it commutes with all the T_i. We are therefore led to the following conclusion. The charged bosons $W^\pm$ supposedly mediating the charged current Fermi interaction (1.6) can be embedded in a gauge theory, which must include the photon in a non trivial way. Furthermore, one must consider also the presence of a further neutral interaction, corresponding to a minimal gauge group $SU(2)XU(1)$. The T_i generate the *weak isospin* $SU(2)$ group, whereas the $U(1)$ group is generated by the hypercharge Y. The two groups are factorized since Y commutes with the T_i. Following Appendix A, it is straightforward to write down the minimal gauge Lagrangian for such a theory. In a concise although also precise notation it is

$$\mathcal{L}_{\min} = -1/4 F^A_{\mu\nu} F^A_{\mu\nu} + i\bar{\Psi}\slashed{D}\Psi \qquad (1.15)$$

where A goes from 1 to $4 = 3+1$ and Ψ includes all the matter fermions in a single column vector of Weyl spinors.

Let us actually be more complete here. We know that each quark occurs in 3 colours, which have not been mentioned so far since they are dummy variables as far as the electroweak interactions are concerned. The 3 colours are there to account for the strong interactions also via a gauge theory with an $SU(3)$ gauge group. Therefore things are such that the Lagrangian in (1.15) can incorporate the strong interactions as well, provided we consider the gauge group $SU(3)XSU(2)XU(1)$ and Ψ properly includes all the fermionic degrees of freedom with the suitable colours. A standard compact notation for Ψ is

$$\Psi = \left(Q(\mathbf{3},\mathbf{2})_{\frac{1}{6}}, L(\mathbf{1},\mathbf{2})_{-\frac{1}{2}}, u^c(\bar{\mathbf{3}},\mathbf{1})_{-\frac{2}{3}}, d^c(\bar{\mathbf{3}},\mathbf{1})_{\frac{1}{3}}, e^c(\mathbf{1},\mathbf{1})_1, N(\mathbf{1},\mathbf{1})_0\right) \tag{1.16}$$

The meaning of it is the following: the two numbers in bold characters denote for any component the $SU(3)$ and the $SU(2)$ representations respectively, whereas the subscript numbers give the hypercharge of each component. For the $SU(2)$-doublets Q and L, the subscript L, for left handed, has been omitted. Notice finally the presence in Ψ of a totally neutral spinor $N(\mathbf{1},\mathbf{1})_0$, normally not included in the Lagrangian of the Standard Model. We are preparing the ground for the neutrino masses. Altogether this makes a total of 16 Weyl spinors, forming a 6 times reducible representation of $SU(3)XSU(2)XU(1)$.

As remarked in the Appendix A, the minimal Lagrangian contains 3 dimensionless couplings, one for each factor of the gauge group. Furthermore, the gauge invariance of this classical action does not fix the hypercharge eigenvalues, which have been chosen to reproduce the observed electric charges of the fermions. This is an embarrassing pending issue to which we shall have to return.

Chapter 2
The Lagrangian of the standard model, including neutrino masses

2.1. The global symmetries of the minimal gauge Lagrangian

The minimal gauge Lagrangian is very far from being realistic. An easy way to see this is to consider its symmetries. The gauge symmetry itself is of course a problem. What distinguishes the photon from the other weak vector bosons? The $SU(2) \times U(1)$ gauge invariance could, at least in principle, be broken by some non perturbative mechanism. There are all reasons to think, in fact, that this would be the case due to the $SU(3)$ sector (Quantum CromoDynamics) but there are also many obvious reasons to think that this would not work. [*Question 2.1.1: Enumerate some of these reasons.*] But a possibly even stronger argument that makes the minimal gauge Lagrangian inadequate is related to its global symmetries.

These global symmetries are tightly related to the high reducibility of the Ψ-multiplet under the full gauge group. As manifest from (1.16), neglecting for the time being the overall singlet $N(\mathbf{1}, \mathbf{1})_0$, Ψ breaks down into 5 irreducible representations of the gauge group, *i.e.* each generator splits into 5 blocks on the diagonal. The same is true, therefore, for the D_μ appearing in the fermionic term of $\mathcal{L}_{\min}$, which splits in a sum of 5 independent pieces

$$i\bar{\Psi}\slashed{D}\Psi = \sum_{\alpha=1}^{5} i\bar{\Psi}_\alpha \slashed{D}_\alpha \Psi_\alpha. \tag{2.1}$$

Hence $\mathcal{L}_{\min}$ is invariant under five independent phase transformations, or five $U(1)$'s. One combination of them is in fact the same hypercharge $U(1)$ gauge group, but the other independent 4 phase transformations must correspond, by Noether's theorem, to 4 globally conserved charges.

What are these charges? It is simple to convince oneself that a possible choice for the appropriate combinations is B, for baryon number, L, for lepton number and, similarly, B_A and L_A for the corresponding symmetries that treat left-handed and right-handed particles in an opposite way (hence with the same *axial* charge for Q and u^c, d^c or for L and

e^c). Now, unlike B and L, the axial B_A and L_A are highly undesirable, at least, but not only, because all quarks and leptons are massive.

This difficulty is in fact exacerbated by the occurrence of the three family replicas of the standard fermions. Now the 5 different blocks in which D_μ splits are actually replicated identically three times each, so that

$$i\bar{\Psi}\slashed{D}\Psi = \sum_{\alpha=1}^{5}\sum_{i=1}^{3} i\bar{\Psi}_\alpha^i \slashed{D}_\alpha \Psi_\alpha^i. \qquad (2.2)$$

With the inclusion of the three families, the global symmetry of $\mathcal{L}_{\min}$ is therefore extended to an overall $SU(3)^5 \times U(1)^4$ global group. Except for B and for the individual family lepton numbers L_e, L_μ, L_τ (we are neglecting neutrino masses for the time being) none of these symmetries is observed in nature.

2.2. Towards a realistic Lagrangian

Keeping the symmetry issue as our guiding line, we ask how one could modify the minimal gauge Lagrangian to turn it into a realistic theory. For good reasons (see Appendix D), we want to stick in the Lagrangian to monomials of dimension 4 at most. Following Appendix A, the only other term that we could possibly add to $\mathcal{L}_{\min}$ is a fermion mass term of the form $\Psi M \Psi$. It is straightforward to see, however, that the entire set of the Ψ fields in equation (1.16), with the exception of the singlet $N(\mathbf{1}, \mathbf{1})_0$, form a chiral representation (see the definition in Appendix B) of the $SU(3) \times SU(2) \times U(1)$ gauge group. [*Problem 2.2.1: Prove this statement.*] The only term that we can add is a mass for the *right-handed neutrinos* (This is how we call the N_i)

$$\Delta \mathcal{L}_M = -\frac{1}{2} N_i M_{ij} N_j + \text{h.c.} \qquad (2.3)$$

where we make explicit also a generation index $i = 1, 2, 3$. M_{ij} in equation (2.3) is an arbitrary symmetric matrix. This term, however, does not break any of the symmetries discussed in Sect. 2.1, since the N_i only enter $\mathcal{L}_{\min}$ through their free kinetic term.

Always following Appendix A, the only way that we have to try to attack this problem is by adding new fields. The way taken by the Standard Model is to add a $SU(2)$ doublet of complex scalars

$$\phi = \begin{pmatrix} \phi^+ \\ \phi^0 \end{pmatrix} \equiv \begin{pmatrix} \phi_1 + i\phi_2 \\ \phi_3 + i\phi_4 \end{pmatrix}, \qquad (2.4)$$

with hypercharge chosen in such a way as to allow its possible Yukawa interactions with the Ψ fields. With a single scalar doublet, this is possible if ϕ transforms as a $(\mathbf{1}, \mathbf{2})_{1/2}$, in which case the most general Yukawa Lagrangian is

$$\Delta \mathcal{L}_Y = -\phi(Q_i \lambda^U_{ij} u^c_j + L_i \lambda^N_{ij} N_j) - \phi^+(Q_i \lambda^D_{ij} d^c_j + L_i \lambda^E_{ij} e^c_j) + \text{h.c.} \quad (2.5)$$

[*Problem 2.2.2: Prove this statement.*] As usual we leave implicit the gauge indices and their contractions. The four 3×3 λ matrices have only an explicit index in family (or flavour) space. Finally another term that we can include in the Lagrangian is a gauge-invariant, hermitian potential in the fields ϕ, ϕ^+, which takes the form

$$V = -\mu^2 \phi^+ \phi + \lambda (\phi^+ \phi)^2, \quad (2.6)$$

with μ^2 and λ real parameters of mass dimension 2 and zero respectively.

2.3. The accidental symmetries of the Standard Model

We have been led in this way to construct the full Lagrangian as

$$\mathcal{L}_{\nu\text{SM}} = \mathcal{L}_{\min} + (\Delta \mathcal{L}_M + \Delta \mathcal{L}_Y) - V, \quad (2.7)$$

where we have included in $\mathcal{L}_{\min}$ also the covariant kinetic term for the doublet ϕ, $|D_\mu \phi|^2$ (see Appendix A). $\mathcal{L}_{\nu\text{SM}}$ has the non trivial property of being the most general Lagrangian, gauge invariant under $SU(3) \times SU(2) \times U(1)$, with monomials of dimension four at most and involving the fermion multiplet Ψ, as in (1.16), and the doublet scalar ϕ. At least, but not only, for historical reasons, it is best to isolate in equation (2.7) the terms which involve the right handed neutrino fields, so that

$$\mathcal{L}_{\nu\text{SM}} = \mathcal{L}_{\text{SM}} + \bar{N}_i \partial\!\!\!/ N_i - \left(\phi L_i \lambda^N_{ij} N_j + \frac{1}{2} N_i M_{ij} N_j + \text{h.c.} \right) \quad (2.8)$$

In the literature one refers to $\mathcal{L}_{\text{SM}}$ as the Standard Model Lagrangian. The experimental discovery of the neutrino masses makes it natural to extend the Standard Model Lagrangian to introduce the right handed neutrinos as well.

Once again: what about the symmetries of this Lagrangian? The gauge symmetry has of course remained intact, since, after all, this has been our very guiding principle. But the potential of the scalar doublet can spontaneously break it in a phenomenologically consistent way. Furthermore, and not less importantly, the global symmetry of $\mathcal{L}_{\min}$ has been drastically reduced. Let us see how, keeping the distinction between $\mathcal{L}_{\text{SM}}$ and $\mathcal{L}_{\nu\text{SM}}$, since, after all, the neutrino masses only introduce small effects.

There is a great redundancy in the various λ matrices in equation (2.5), which it is useful to reduce away. To this end, we first diagonalize them by appropriate unitary rotations

$$\lambda^I = (V_L^I)^T \lambda_d^I V_R^I \qquad I = U, N, D, E \tag{2.9}$$

where λ_d^I are real diagonal matrices with non negative eigenvalues. Then we notice that, by appropriate unitary transformations of the different components of the Ψ multiplet that do not affect $\mathcal{L}_{\min}$, $\Delta\mathcal{L}_Y$ can be reduced to

$$\Delta\mathcal{L}_Y \Rightarrow -\phi(Q^T V_q^T \lambda_d^U u^c + L^T V_l^T \lambda_d^N N)$$
$$-\phi^+(Q^T \lambda_d^D d^c + L^T \lambda_d^E e^c) + \text{h.c.} \tag{2.10}$$

where V_q and V_l are again unitary matrices related to the V_L^I as $V_q = V_L^U(V_L^D)^+$ and $V_l = V_L^N(V_L^E)^+$. We have also not bothered renaming the transformed fields. Similarly, in the transformed basis for the right handed neutrino fields, the mass matrix M changes, but we shall keep for it the same name without any loss of information. [*Problem 2.3.1: Perform explicitly this transformation.*]

To summarize, we have reduced the Standard Model Lagrangian to

$$\mathcal{L}_{\text{SM}} = \mathcal{L}_{\min} - (\phi Q^T V_q^T \lambda_d^U u^c + \phi^+ Q^T \lambda_d^D d^c + \phi^+ L^T \lambda_d^E e^c + \text{h.c.}) - V \tag{2.11}$$

and the total Lagrangian to

$$\mathcal{L}_{\nu\text{SM}} = \mathcal{L}_{\text{SM}} + \bar{N}\slashed{\partial}N - \left(\phi L^T V_l^T \lambda_d^N N + \frac{1}{2}N^T M N + \text{h.c.}\right). \tag{2.12}$$

In this way the global symmetries are made manifest. In the quark sector of $\mathcal{L}_{\text{SM}}$, due to the presence of V_q, only the overall baryon number is respected, whereas in the lepton sector, again in $\mathcal{L}_{\text{SM}}$, the individual family lepton numbers, L_e, L_μ and L_τ are conserved. This is not the least achievement of the Standard Model, given the experimental status of these charges, as we shall see later on. When the right-handed neutrinos and hence the neutrino masses are added, their Yukawa interactions break L_e, L_μ and L_τ, while conserving the overall lepton number, which is broken by the right-handed neutrino mass terms. $\mathcal{L}_{\nu\text{SM}}$ therefore becomes a candidate for a fully realistic theory of the electroweak interactions.

Chapter 3
The main predictions of the standard model

3.1. Gauge symmetry breaking and particle masses

If μ^2 in equation (2.6) is positive, which we assume hereafter, the potential V has a minimum at

$$\phi^+\phi = \frac{\mu^2}{2\lambda} \equiv v^2. \tag{3.1}$$

Also for later purposes, it is useful to notice that $\phi^+\phi = \phi_1^2+\phi_2^2+\phi_3^2+\phi_4^2$, so that the potential V has actually a $SO(4)$ invariance, of which the gauged $SU(2)XU(1)$ is a subgroup.

Following the general discussion in Appendix C, it is therefore possible to choose a particular vacuum configuration, where

$$\langle\phi\rangle = \begin{pmatrix} 0 \\ v \end{pmatrix} \tag{3.2}$$

and v is real. This leaves a single generator unbroken, $Q = T_3 + Y$, and a corresponding single massless vector, the photon, whereas the 3 remaining vectors become massive. Their mass Lagrangian, upon insertion of (3.2) into the covariant kinetic term of the ϕ bosons, is

$$\Delta\mathcal{L}_m = \frac{1}{2}\frac{v^2}{2}\left[g^2(W_\mu^1)^2 + g^2(W_\mu^2)^2 + (-gW_\mu^3 + g'B_\mu)^2\right], \tag{3.3}$$

where W_μ^i and B_μ are the $SU(2)$ and $U(1)$ gauge bosons respectively, and g, g' the corresponding couplings. We then have a charged vector boson $W_\mu^\pm = 1/\sqrt{2}(W_\mu^1 \pm W_\mu^2)$ of mass

$$m_W = gv/\sqrt{2} \tag{3.4}$$

or, from equation (1.7),

$$G_F^{-1} = 2\sqrt{2}v^2. \tag{3.5}$$

Similarly a neutral boson,

$$Z_\mu = \cos\theta W_\mu^3 - \sin\theta B_\mu, \tag{3.6}$$

with mixing angle

$$\theta = \arctan\frac{g'}{g}, \tag{3.7}$$

acquires the mass

$$m_Z = \frac{\sqrt{g^2 + g'^2}}{\sqrt{2}} v, \tag{3.8}$$

whereas the photon is the ortogonal massless combination

$$A_\mu = \sin\theta W_\mu^3 + \cos\theta B_\mu. \tag{3.9}$$

Notice the relation between the masses of the charged and neutral bosons,

$$\frac{m_W^2}{m_Z^2 \cos^2\theta} \equiv \rho = 1. \tag{3.10}$$

They become in particular degenerate in the $g' \to 0$ limit or $\cos\theta \to 1$. This is not a consequence of the gauge symmetry but rather of the $SO(4)$ symmetry of the potential, as we now show. [*Problem* 3.1.1: *Determine the ratio of the charged to the neutral W-boson masses, if the SU(2) symmetry is broken by the vacuum expectation value of a scalar SU(2)-triplet with zero hypercharge.*]

$SO(4)$ is isomorphic to $SU(2) \times SU(2)$. In fact, defining the 2x2 matrix

$$\mathcal{H} = (i\sigma_2 \phi^*, \phi), \tag{3.11}$$

its action on ϕ can be identified as

$$\mathcal{H} \Rightarrow e^{i\omega_L^i \frac{\sigma_i}{2}} \mathcal{H} e^{-i\omega_R^i \frac{\sigma_i}{2}}, \tag{3.12}$$

where $\omega_{L,R}^i$ are the parameters of two independent $SU(2)$ transformations, often referred to as $SU(2)_L \times SU(2)_R$. $SU(2)_L$ actually coincides with the $SU(2)$ of the gauge group. This gauging respects therefore the full $SU(2)_L \times SU(2)_R$, with the W_μ^i transforming as a triplet under $SU(2)_L$, whereas the gauging of the hypercharge $U(1)$ does not. Since the vacuum configuration $\langle \mathcal{H} \rangle = v\mathbf{1}$ is left invariant by the *diagonal* $SU(2)_V$ with $\omega_L^i = \omega_R^i$ in (3.12), this explains the degeneracy of the neutral and the charged vector boson masses as $g' \to 0$. Being responsible for keeping $\rho = 1$, $SU(2)_L \times SU(2)_R$ is often called *custodial* symmetry. [*Problem* 3.1.2: *Write the Lagrangian of the Standard Model in terms of*

$\mathcal{H}$ rather than ϕ. *Question 3.1.1: Which is the dominant coupling that breaks the $SU(2)_L \times SU(2)_R$ symmetry other than gı? Problem 3.1.3: Show that a $U(1)$ transformation on ϕ is equivalent to a right multiplication of $\mathcal{H}$ by $\exp(-i\omega_R^3 \sigma_3/2)$.*]

Like the gauge bosons, also the fermions acquire a mass after spontaneous symmetry breaking of the gauge symmetry. From equation (2.10), replacing ϕ with its vacuum expectation value, the mass eigenvalues of all the charged fermions are obtained:

$$m_i^U = (\lambda_d^U)_{ii} v, \quad m_i^D = (\lambda_d^D)_{ii} v, \quad m_i^E = (\lambda_d^E)_{ii} v. \tag{3.13}$$

In the case of the up-type quarks, to go to the basis of the mass eigenstates or the *physical* basis, we have redefined the left handed components as

$$u_i \rightarrow (V_q)^*_{ji} u_j. \tag{3.14}$$

The neutrino masses will be described in Lecture 7.

3.2. Couplings to fermions of the gauge bosons

The couplings of the vector bosons to a generic fermion arise from the covariant derivative

$$D_\mu = \partial_\mu - ig W_\mu^i T^i - ig' Y B_\mu, \tag{3.15}$$

which, after introducing the physical vectors, becomes

$$\begin{aligned} D_\mu = \partial_\mu &- i\frac{g}{\sqrt{2}}(W_\mu^+ T^- + W_\mu^- T^+) \\ &- i\frac{g}{\cos\theta} Z_\mu (T^3 - \sin^2\theta Q) - i\frac{g}{\sin\theta} A_\mu Q. \end{aligned} \tag{3.16}$$

The two parameters g, g' that determine all these couplings, can also be traded for the electromagnetic coupling and for the mixing angle

$$e = g\sin\theta, \quad \theta = \arctan\frac{g'}{g}. \tag{3.17}$$

[*Problem 3.2.1: Write down the neutral current interacting with the Z-boson in terms of all the explicit fermion fields.*]

3.3. The Higgs boson

As discussed in general in Appendix C, a useful parametrization of the doublet ϕ is

$$\phi(x) = e^{i\pi_i(x)\sigma_i/v} \begin{pmatrix} 0 \\ v + \frac{h(x)}{\sqrt{2}} \end{pmatrix}, \tag{3.18}$$

where the π_i are the Goldstone bosons eaten up by the W and the Z, whereas $h(x)$ is the single physical scalar remaining in the spectrum, the Higgs boson. Getting rid of the π_i (*i.e.* going to the unitary gauge) and substituting (3.18) into the potential (2.6) gives

$$V = \frac{1}{2}m_h^2 h^2 + \sqrt{\frac{\lambda}{2}} m_h h^3 + \frac{\lambda}{4} h^4 \qquad (3.19)$$

where

$$m_h = 2\sqrt{\lambda}v = \sqrt{2}\mu. \qquad (3.20)$$

In this way the two parameters of the original Higgs potential, μ^2 and λ, determine the Fermi scale, equation (3.5), the physical Higgs mass and the Higgs self-interactions. From loop corrections, the coupling λ becomes energy dependent and increases at high energies. Therefore (3.19) is perturbatively predictive only up to a maximal energy E_{max} dependent on m_h: $E_{max} \approx 2$ TeV for $m_h \approx 600$ GeV, up to $E_{max} \approx 10^{18}$ GeV for $m_h \approx 180$ GeV. [*Problem 3.3.1: Calculate the energy dependence of the one loop correction to λ from the self-coupling itself.*] At the time of writing these lectures, the searches of the Higgs boson in e^+e^- collisions at LEP have only produced a lower bound on its mass, $m_h > 114$ GeV.

Proceeding in a similar way, one obtains the couplings of the Higgs boson to the vectors:

$$\mathcal{L}_{h-V} = \left[m_W^2 W_\mu^+ W_\mu^- + \frac{1}{2} m_Z^2 Z_\mu Z_\mu \right] \left(1 + \frac{1}{\sqrt{2}v} h \right)^2 \qquad (3.21)$$

and to the charged fermions

$$\mathcal{L}_{h-f} = -\Sigma_f m_f \bar{f}_D f_D \left(1 + \frac{1}{\sqrt{2}v} h \right) \qquad (3.22)$$

where, for every fermion of mass m_f, we have introduced a Dirac spinor $f_D = f + (f^c)^C$ and C stands for the charge conjugation operation.

Chapter 4
Precision tests

The number of precision measurements that test the consequences of $\mathcal{L}_{SM}$ is overwhelming. The prototype examples are of course the electron or muon anomalous magnetic moments. These tests explore the theory in a very wide range of scales, from the tiniest frequencies in atomic physics up to detailed properties of the Z boson. In the following we shall be concerned only with a few of these tests. The selection criterion is somewhat arbitrary. We focus on those examples which not only test the weak sector of the Standard Model Lagrangian to a high level of accuracy but (we think) are also more likely to give indirect information on new physics perhaps hiding behind the Standard Model.

4.1. Parity violation in atomic physics

Parity violation in atoms is a precisely measured manifestation of the weak interactions at the smallest energy scales. It has also been one of the very first manifestations of parity violation in the neutral current interactions (or the interactions of the Z boson).

On general symmetry grounds, a parity violating non relativistic Hamiltonian describing a contact interaction between the electron and the nucleus, neglecting nuclear spin, must have the form

$$\mathcal{H}_{PV} = Q_W \frac{G_F}{4\sqrt{2} m_e} \vec{\sigma}_e \cdot [\vec{p}_e \delta^{(3)}(\vec{r}_e) + \delta^{(3)}(\vec{r}_e) \vec{p}_e], \quad (4.1)$$

where $(\vec{r}, \vec{p}, \vec{\sigma})_e$ are the coordinate, the momentum and the spin of the electron, m_e its mass and Q_W is an unknown dimensionless coefficient, called *weak charge* of the atom.

To express Q_W in terms of the fundamental couplings, we have to start from the effective electron-quark interaction arising from the Z-exchange

$$\mathcal{L}_{PV}^{eff} = \frac{G_F}{\sqrt{2}} \sum_{q=u,d} [c_{1q}(\bar{e}_D \gamma_\mu \gamma_5 e_D)(\bar{q}_D \gamma_\mu q_D) + c_{2q}(\bar{e}_D \gamma_\mu e_D)(\bar{q}_D \gamma_\mu \gamma_5 q_D)], \quad (4.2)$$

where we only keep the parity odd terms. The numerical coefficients c_{1q} and c_{2q} are obtained from (3.16)

$$c_{1u} = -\frac{1}{2} + \frac{4}{3}\sin^2\theta,$$
$$c_{1d} = \frac{1}{2} - \frac{2}{3}\sin^2\theta, \qquad (4.3)$$
$$c_{2u} = -c_{2d} = -\frac{1}{2} + 2\sin^2\theta.$$

[*Problem 4.1.1: Prove these equations.*]

In the non relativistic limit for the electron and in the static point-like limit for the nucleus, it is only the $\mu = 0$ component of the first term in the r.h.s. of equation (4.2) that contributes, giving in coordinate space equation (4.1) with

$$Q_W = -2(c_{1u}n_u + c_{1d}n_d), \qquad (4.4)$$

where $n_{u,d}$ are the occupation numbers of the u, d quarks in the nucleous. [*Problem 4.1.2: Prove equation (4.4).*] In turn these occupation numbers are linear combinations of baryon number and electric charge, hence of the number of protons Z and neutrons N in the atomic nucleus,

$$n_u = 2Z + N, \qquad n_d = Z + 2N. \qquad (4.5)$$

This completes the determination of Q_W in terms of the fundamental parameters of the theory.

Although the matrix elements of (4.1) for a heavy atom are roughly proportional to an important Z^3 factor, their absolute value is still very small and must be compensated, to see an effect, by a small energy denominator. This is the case in the actual measurements, where one effectively looks at optical transitions from a pair of quasi-degenerate states of opposite parity, mixed by $\mathcal{H}_{PV}$, to a lower state of definite parity. By measurements on cesium atoms, which have $Z = 55$ and $N = 78$, the weak charge $Q_W(Cs) \approx -72$ is obtained in good agreement with the theoretical expectation for a reference value of the weak mixing angle, as determined elsewhere. The relative error of this comparison is of about 5 ppm, dominated by the uncertain knowledge, in the extraction of $Q_W(Cs)$ from the experiment, of the relevant matrix element of $\mathcal{H}_{PV}$ in the complex multi-electron atom.

4.2. Leading corrections to the ρ parameter

The ρ parameter, as defined in equation (3.10), is a key quantity in the context of the ElectroWeak Precision Tests, since it is measured with high accuracy, better than 1 ppm, and may be especially affected by new phenomena. At the beginning of the the 90's, before the direct discovery of the top quark at Fermilab, the ρ parameter has provided the main indirect information on the top mass, m_t, which was correctly determined with about 30% accuracy. With current data, the indirect determination of the top mass is at the few % level and agrees with the direct measurement, $m_t = 171.4 \pm 2.1$ GeV. This is possible at all because the tree level relation, $\rho = 1$, receives corrections from the couplings that break the custodial $SU(2)_L \times SU(2)_R$ symmetry described in Sec. 3.1. In the Standard Model the dominant such coupling is the Yukawa coupling of the top quark, hence the top quark mass (see Question 3.1.1).

Since these top-quark corrections to the ρ parameter have nothing to do with the gauge symmetry, (as the very same tree level relation $\rho = 1$), it is of interest to see them in the *gauge-less* limit of the Standard Model, where the gauge bosons are external non propagating fields. The reference Lagrangian that we need to consider is therefore

$$\mathcal{L}_{\text{gauge-less}} = i\bar{\Psi}\slashed{D}\Psi + |D_\mu \phi|^2 - V(\phi) - (\lambda_t \phi Q_3 t^c + h.c.) \quad (4.6)$$

where Ψ only includes the third-generation fermions, the doublet Q_3 and the singlet t_c, ϕ is the Higgs doublet and λ_t is the top Yukawa coupling. The covariant derivatives contain the gauge bosons fields, which do not have, however, a kinetic term.

It is useful to see the role of the eaten up Goldstone bosons, π_i, in this Lagrangian. To this end let us insert in it equation (3.18). Expanding to leading order in the π_i's, we get their covariant kinetic terms

$$\mathcal{L}_{\text{kin}}(\pi_i) = Z_2^{(+)} \left|\partial_\mu \pi^+ - g\frac{v}{\sqrt{2}} W_\mu^+\right|^2 + \frac{Z_2^{(0)}}{2}\left(\partial_\mu \pi^0 - \frac{gv}{2\cos\theta} Z_\mu\right)^2 \quad (4.7)$$

where $\pi^\pm = (\pi_1 \pm i\pi_2)/\sqrt{2}$ and $\pi^0 = \pi_3$. We have also inserted two wave-function renormalization constants $Z_2^{(+)}$ and $Z_2^{(0)}$ for π^+ and π^0 respectively, which are both 1 to lowest order but will deviate from 1 at higher orders in perturbation theory, while keeping however the same covariant form of the kinetic terms of the π's. Expanding the squares in (4.7) gives an all order result, in the gauge-less limit, for the ρ parameter

$$\rho = \frac{m_W^2}{m_Z^2 \cos^2\theta} = \frac{Z_2^{(+)}}{Z_2^{(0)}}, \quad (4.8)$$

in terms of the ratio of the wave function renormalization constants for the eaten-up Goldstone bosons.

This gives an effective way to calculate the deviation from 1 of the ρ parameter arising from the top Yukawa coupling. Notice that it is only the ratio of the wave function renormalization constants which is finite in the ultraviolet. At one loop order, from the diagrams of Figure 4.1,

$$\rho = 1 + \frac{3\lambda_t^2}{32\pi^2} = 1 + \frac{3G_F m_t^2}{8\sqrt{2}\pi^2}, \qquad (4.9)$$

i.e. almost a 1% correction. [*Problem 4.2.1: Compute this correction using (4.8).*]

Along similar lines it can be shown that the only other observable receiving one loop corrections proportional to m_t^2 is the Z-width into a $b\bar{b}$ pair. In fact the interaction of the Z-boson with the left handed component of the b-quark gets modified to

$$i\frac{g}{\cos\theta}\left(\frac{1}{2} - \frac{1}{3}\sin^2\theta + \frac{1}{2}\tau\right) Z_\mu \bar{b}_L \gamma_\mu b_L, \quad \tau = -\frac{G_F m_t^2}{4\pi^2\sqrt{2}}, \qquad (4.10)$$

as it can again be easily computed in the *gauge-less* limit by working out the one-loop derivative coupling of π^0 to b_L.

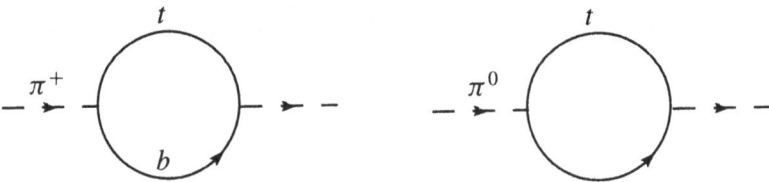

Figure 4.1. One loop corrections to the propagators of the Goldstone bosons from top-bottom exchanges.

4.3. Sensitivity to the Higgs mass

Is it not possible, like it has been for the top quark, to get an indirect information from the precision tests also on the Higgs boson mass? The answer is yes, but the sensitivity on m_h is far less important than the one on m_t. Once again this goes back to the $SU(2)_L \times SU(2)_R$ symmetry, that, as we saw, is exactly respected by the Higgs potential. As such, there cannot be any one loop corrections to ρ proportional to λ, the quartic Higgs coupling, which would mean corrections growing like m_h^2. To find such type of corrections one has to go to two loops, so that the necessary breaking of the $SU(2)_L \times SU(2)_R$ symmetry is allowed to come in. These corrections, for $m_h < 1$ TeV, are too small to be of any interest.

There exist, however, significant one loop corrections growing like the logarithm of m_h. They affect 6 independent amplitudes, all involving external vector boson lines only: two each for the 2, 3 and 4 point functions. Only the 2 point functions, or vacuum polarization amplitudes, are experimentally significantly constrained (see below). We then concentrate on these ones.

Let us define 4 vacuum polarization amplitudes in an effective Lagrangian language, after breaking of the gauge symmetry and including the one loop corrections, in momentum space

$$\mathcal{L}_{\text{vac-pol}} = -\frac{1}{2} W_\mu^3 \Pi_{33}(q^2) W_\mu^3 - \frac{1}{2} B_\mu \Pi_{00}(q^2) B_\mu \\ - W_\mu^3 \Pi_{30}(q^2) B_\mu - W_\mu^+ \Pi_{WW}(q^2) W_\mu^-. \qquad (4.11)$$

We neglect the *longitudinal* terms, proportional to $(q_\mu A_\mu)^2$, where A_μ is any vector, since they are irrelevant for physical amplitudes with external fermion lines. [*Question 4.3.1: Why?*] These amplitudes are in general gauge dependent. Here we deal only with those effects that are gauge independent, like the coefficient of the $\log m_h$ terms. Expanding these amplitudes in q^2, by power counting only the $\Pi(0)$'s and the $\Pi'(0)$'s can diverge as $m_h \to \infty$. There are 8 such terms. Three of them can be traded for the three parameters which enter at tree level: v, g, g'. The masslessness of the photon entails two conditions on the $\Pi(0)$'s. There remain therefore three effective parameters predicted in the SM and potentially growing with m_h. Let us see all this more in detail.

We normalize the W_μ^i and the B_μ by setting respectively $\Pi'_{WW}(0)$ and $\Pi'_{00}(0)$ to 1. This defines g, g' through the couplings to fermions. We are therefore left with two predicted $\Pi'(0)$'s which we take as

$$\hat{S} = \frac{g}{g'} \Pi'_{30}(0), \quad \hat{U} = \Pi'_{33}(0) - \Pi'_{WW}(0). \qquad (4.12)$$

On the $\Pi(0)$'s, the zero mass of the photon requires

$$\Pi_{\gamma\gamma}(0) = \Pi_{\gamma Z}(0) = 0, \qquad (4.13)$$

which, from (3.9) and (3.6), are equivalent to

$$\frac{\Pi_{33}(0)}{\cos^2 \theta} = \frac{\Pi_{00}(0)}{\sin^2 \theta} = -\frac{\Pi_{30}(0)}{\sin \theta \cos \theta} \equiv \Pi_{ZZ}(0), \qquad (4.14)$$

determined in terms of v. This leaves us with a single predicted term

$$\hat{T} = \frac{\Pi_{33}(0) - \Pi_{WW}(0)}{m_W^2}. \qquad (4.15)$$

Note that both $\hat{T}$ and $\hat{U}$ break the custodial symmetry. As we already said, this forbids one loop terms growing like m_h^2 which, by power counting, could be present in $\hat{T}$ but are instead converted into a $\log m_h$. Similarly the derivative terms may have a $\log m_h$ by power counting, but this is only present in $\hat{S}$, which is non vanishing in the $g' \to 0$ limit. $\hat{S}$ is associated with first order breaking of $SU(2)_L \times SU(2)_R$ by the hypercharge gauge coupling, which is however absorbed away by the explicit g'-factor in the denominator of $\hat{S} = (g/g')\Pi'_{30}(0)$. [*Problem 4.3.1: Calculate the four amplitudes in equation (4.11) at tree level in the Standard Model and the corresponding value of $\hat{S}$, $\hat{T}$ and $\hat{U}$.*]

One way to compute the coefficients of the $\log m_h$ terms for $\hat{S}$ and $\hat{T}$ is to view m_h as the cut-off of the divergent vacuum polarization diagrams where there is no Higgs boson as an internal line. In this way one gets

$$\hat{S} \approx \frac{G_F m_W^2}{12\sqrt{2}\pi^2} \log m_h, \qquad \hat{T} \approx -\frac{3 G_F m_W^2}{4\sqrt{2}\pi^2} \tan^2\theta \log m_h. \qquad (4.16)$$

[*Problem 4.3.2: Show that the result for $\hat{S}$ can be reproduced by calculating the divergence of the diagram of Figure 4.2a, where the internal lines are the charged Goldstone bosons, propagating in any ξ-gauge.*]

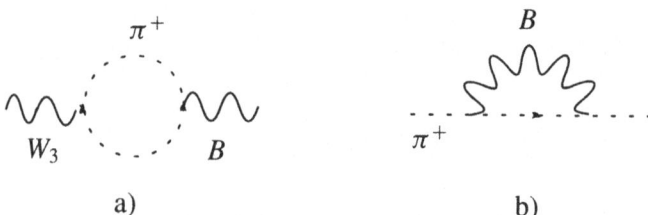

a)   b)

Figure 4.2. a) One loop contribution to $\hat{S}$ from Goldstone boson exchanges; b) One loop correction from B exchange to the propagator of the charged Goldstone boson.

As anticipated, these effects serve to bound experimentally the Higgs boson mass in the Standard Model, since $\hat{S}$ and $\hat{T}$ affect all the precision observable in a definite way. [*Problem 4.3.3: Show that $\hat{T}$ affects the ρ parameter as $\rho - 1 = \hat{T}$. Problem 4.3.4: In the Landau gauge, where the propagating Goldstone bosons are massless, use equation (4.8) to show the result for $\hat{T}$ in (4.16) by calculating the divergence of the diagram of Figure 4.2b.*] Figure 4.3, from the analysis of the data at the time of writing these lectures, shows this constraint by comparing the experimental determination of $\hat{S}$ and $\hat{T}$ with the prediction in the Standard Model as function of m_h. The reference point $\hat{S} = \hat{T} = 0$ is conventionally taken to correspond to the Standard Model value of $\hat{S}$ and $\hat{T}$ at $m_h = 115$ GeV and $m_t = 175$ GeV. Therefore what the figure shows is the possibly required deviation from such reference value. In fact one can forget about this reference value and view the figure as the required deviation of $\hat{S}$ and $\hat{T}$ from the prediction of the Standard Model, shown for $m_t = 171.4$ GeV, the current central value of the latest direct determination of the top quark mass, and m_h varying between 100 and 500 GeV. Since the relevant m_h-region turns out to be relatively low, close to the Z mass, an accurate fit requires including also terms that vanish in the large m_h limit, which explains the slight bending of the theoretical curve for increasing m_h. From the full fit of the ElectroWeak Precision Tests in the Standard Model one obtains at present the indirect determination

$$m_h = 85^{+39}_{-28} \text{ GeV}, \quad m_h < 165 \text{ GeV at } 95\% \text{ CL}. \qquad (4.17)$$

This upper bound on m_h apparently stronger than the one readable from Figure 4.3 is due to the correlation between $\hat{S}$ and $\hat{T}$ in term of a single parameter m_h, valid in the Standard Model, which increases the number of degrees of freedom of the Standard Model fit.

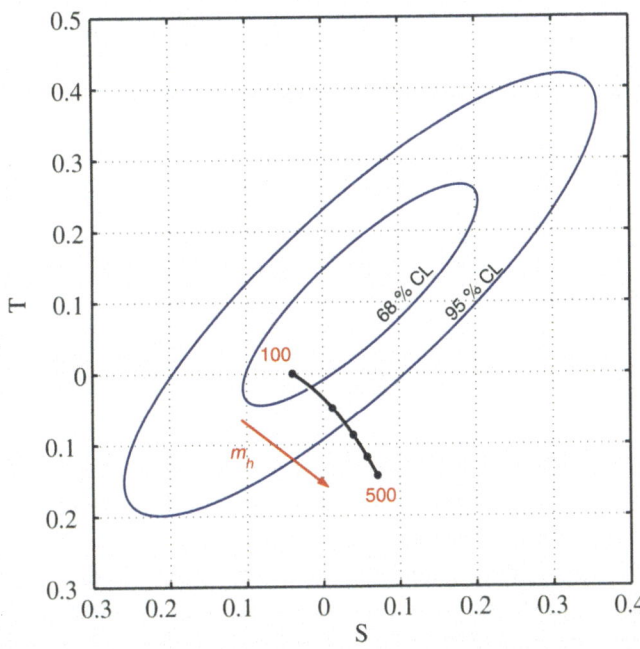

Figure 4.3. $\hat{S}$ and $\hat{T}$ as experimentally determined, compared with the prediction of the Standard Model as function of the Higgs boson mass in GeV and $m_t = 171.4$ GeV. As frequently done in the literature, the axes are scaled to $S \equiv \frac{4\sin\theta^2}{\alpha}\hat{S} \approx 120\hat{S}$ and $T \equiv \frac{1}{\alpha}\hat{T} \approx 130\hat{T}$.

4.4. Vacuum polarization amplitudes in a general *universal* theory

All the above considerations apply to the Standard Model and give rise to a significant, although indirect, constraint on the Higgs boson mass. What if some new physics beyond the Standard Model were hiding in the ElectroWeak Precision Tests? To answer this question in full generality, without any commitment to the form of this putative new physics, is hardly possible at all. It is possible, on the contrary, to make some useful considerations, if one restricts oneself to *universal* theories, *i.e.* theories where deviations from the Standard Model, to leading order, only appear through the vector bosons vacuum polarizations in (4.11). Furthermore we assume that the characteristic scale, Λ_{NP}, associated with the new physics be large enough that it is meaningful to expand the vacuum polarization amplitudes in powers of q^2. The effective parameters $\hat{S}$, $\hat{T}$ and $\hat{U}$ continue to serve to the purpose of this Section. We consider however also the second derivative terms in the expansion

$$\Pi_V(q^2) \simeq \Pi_V(0) + q^2 \Pi'_V(0) + \frac{(q^2)^2}{2!}\Pi''_V(0), \qquad (4.18)$$

in terms of the new dimensionless effective parameters:

$$\begin{aligned}
W &= \frac{m_W^2}{2}\Pi''_{33}, \\
Y &= \frac{m_W^2}{2}\Pi''_{00}, \\
V &= \frac{m_W^2}{2}(\Pi''_{33} - \Pi''_{WW}), \\
X &= \frac{m_W^2}{2}\Pi''_{30},
\end{aligned} \qquad (4.19)$$

all at $q^2 = 0$. Needless to say, all these parameters are predicted in the Standard Model, where they vanish at tree level. Here we deal with the possible deviations from them.

In total we are dealing with 7 effective parameters. In order to decide their relative importance, it is useful to recall the consistency of every form factor with gauged $SU(2)_L$ and with the custodial symmetry (as $g' \to 0$), the two relevant symmetries in the problem, summarized in Table 4.1. Within each set with the same symmetry properties, it is then natural to retain the form factor of relative lowest order in the number of derivatives, since, barring accidental cancellations, the other terms will give effects suppressed by a factor of q^2/Λ_{NP}^2. This leads to the emergence of four effective form factors: $\hat{S}$, $\hat{T}$, W and Y.

Table 4.1. Symmetry properties of the various form factors. "+" means "symmetric", "−" means "non-symmetric". Every form factor can be non zero even in absence (+) or only in presence (−) of a breaking of the corresponding symmetry (as $g' \to 0$).

Form factor	$\hat{S}$	$\hat{T}$	$\hat{U}$	V	X	W	Y
custodial	+	−	−	−	+	+	+
gauged $SU(2)_L$	−	−	−	−	−	+	+

4.5. Current experimental constraints

The main constraints on these effective form factors arise from two sets of measurements:

1. The various observables at the Z-pole in e^+e^- annihilation and the W-mass;
2. The cross-sections and the asymmetries in $e^+e^- \to f\bar{f}$ at the highest center of mass energies reached at LEP2, or $q^2 \approx (200 \text{ GeV})^2$.

On general grounds it can be shown that the first class of measurements, with typical 1 ppm precision, constrain three combinations of these form factors. Hence the LEP2 measurements, of about 1% precision, play a crucial role in determining the whole set. Notice that the lower precision of LEP2 is almost compensated by the higher center of mass energy, which enhances the effect of W and Y. The bounds on $\hat{S}$, $\hat{T}$, W and Y from a global fit to all these data are shown in Table 4.2.

Table 4.2. Global fit of possible extra effects in the dominant form factors including them one-by-one or all together, relative to the Standard Model values with a light ($m_h = 115$ GeV) and with a heavy ($m_h = 800$ GeV) Higgs boson.

Type of fit	$10^3 \hat{S}$	$10^3 \hat{T}$	$10^3 Y$	$10^3 W$
One-by-one (light Higgs)	-0.1 ± 0.6	-0.1 ± 0.6	0.0 ± 0.6	0.2 ± 0.6
One-by-one (heavy Higgs)	—	2.4 ± 0.6	—	—
All together (light Higgs)	-0.7 ± 1.3	-0.5 ± 0.9	-0.4 ± 1.2	0.2 ± 0.8
All together (heavy Higgs)	-1.7 ± 1.3	1.4 ± 1.0	-0.5 ± 1.2	0.3 ± 0.8

As already said, $\hat{S}$, $\hat{T}$, W and Y represent the deviations from the Standard Model. As such the results in Table 4.2 depend on the Higgs boson mass, which influences in a significant way the first class of measurements. Table 4.2 shows that a non zero $\hat{T}$ from new physics can in principle compensate the effect of a heavy Higgs boson mass. This may be a technical accident or it may mean that some new physical effect could actually be hidden in the ElectroWeak Precision Tests, thus confusing the interpretation of the data in terms of the Standard Model only.

4.6. An interlude: making it without a Higgs boson

In all what has been said so far, as it will be the case in the next Lecture about flavour physics, the Higgs doublet plays a crucial role, to the point that one must consider the Higgs boson as an intrinsic component of the Standard Model. While this is true, we should perhaps not forget that no direct experimental evidence for it has emerged so far. There is a clear reason for this: no experiment has had enough sensitivity to explore the entire relevant mass range for the Higgs. At the same time, the very same consistency of the Standard Model with the EWPT speaks significantly, although indirectly, in favour of the existence of a Higgs boson. Nevertheless it remains at least logically valid to ask how far one can go in describing ElectroWeak physics without introducing a Higgs boson at all.

For the consistency of the theory, we cannot renounce gauge invariance, which therefore must be spontaneously broken. Discussing how this could happen is outside the scope of these Lectures. We can nevertheless try to characterize in a general way some features of the resulting physical theory. The key question is how to describe the *massive* vector bosons and the *massive* fermions of the Standard Model in a fully gauge invariant way *without* introducing any extra degree of freedom, like the Higgs boson. For pedagogical reasons, let us see first how this is possible at all by considering a suitable manipulation of the Standard Model itself.

The starting point is the matrix representation (3.11) of the Higgs doublet, introduced in Lecture 3, which, inspired by (3.18), we reparametrize as

$$\mathcal{H} = \left(v + \frac{h}{\sqrt{2}} \right) \Sigma, \quad \Sigma \equiv e^{i\pi_i(x)\sigma_i/v}. \tag{4.20}$$

Remember that the π_i are the Goldstone bosons eaten up by the W and the Z, so, at the end, they will not enter in the physical spectrum. The important point is that the $SU(2) \times U(1)$ gauge symmetry, in fact the full $SU(2)_L \times SU(2)_R$ introduced in association with $\mathcal{H}$, can now be viewed as acting only on Σ and not on h as

$$\Sigma \Rightarrow e^{i\omega_L^i \frac{\sigma_i}{2}} \Sigma e^{-i\omega_R^i \frac{\sigma_i}{2}}. \tag{4.21}$$

Therefore we can write down a fully gauge invariant Lagrangian without even having to mention the field h. The Lagrangian of the Standard Model written in terms of $\mathcal{H}$ (see Problem 3.1.2) with all the h-dependent terms thrown away is itself an example. Needless to say, this Lagrangian describes massive W an Z bosons together with a massless photon.

It is in fact useful to have at our disposal a covariant formalism for writing a general $SU(2) \times U(1)$ invariant Lagrangian only involving the vector bosons and the Σ field (and the fermions, if one wants too). The obvious ingredients of such formalism are:

- The usual field strengths: $W^a_{\mu\nu} \frac{\sigma^a}{2} \equiv \hat{W}_{\mu\nu}$ and $B_{\mu\nu}$;
- The Lorentz vector $V_\mu \equiv (D_\mu \Sigma)\Sigma^+$, involving the covariant derivative of the Σ-field

$$D_\mu \Sigma \equiv \partial_\mu \Sigma + i\frac{g}{2} W^a_\mu \sigma^a \Sigma - i\frac{g'}{2}\Sigma B_\mu \sigma^3; \qquad (4.22)$$

- The combination of Σ-fields $T \equiv \Sigma \sigma^3 \Sigma^+$.

Except $B_{\mu\nu}$ which is an invariant, all these combinations of fields, collectively denoted by Φ, transform under $SU(2)_L$ as

$$\Phi \Rightarrow e^{i\omega^i_L \frac{\sigma_i}{2}} \Phi \, e^{-i\omega^i_L \frac{\sigma_i}{2}}. \qquad (4.23)$$

In turn an arbitrary Lagrangian term invariant under the full gauge symmetry is made of any Lorentz invariant product of Φ's which is invariant under $SU(2)_L$ (and is non trivial, unlike the trace of any product of T's since $\Sigma^+ \Sigma = 1$).

Which of these terms should we include in the Lagrangian? This is a meaningful question since, with Σ dimensionless, the usual dimensionality criterium does not work. We make *an expansion in momenta* or more properly in the number of momentum factors and/or of vector boson legs. Up to second order and with the inclusion of the usual gauge boson kinetic terms, the *chiral ElectroWeak* Lagrangian is

$$\mathcal{L}_{\text{EWCh}} = -\frac{1}{2} Tr(\hat{W}_{\mu\nu} \hat{W}_{\mu\nu}) - \frac{1}{4} B_{\mu\nu} B_{\mu\nu} \\ + \frac{v^2}{4} Tr[(D_\mu \Sigma)^+ D_\mu \Sigma] + a_0 \frac{v^2}{4} [Tr(T V_\mu)]^2 \qquad (4.24)$$

The first three terms are normalized in such a way that the kinetic terms of the W, the B and of the π-fields, included in the expansion of the exponent of the Σ field, are canonical. The last term, which contributes to the masses of the vector bosons, has an arbitrary coefficient, a_0, as are arbitrary the coefficients of all the extra 9 terms that enter at next order in the momentum expansion. [*Problem 4.6.1: Write down the two such terms that have at the same time two external boson legs and put them in correspondence, together with the last term in (4.24), with the form factors $\hat{S}, \hat{T}, \hat{U}$ previously defined.*] All of these coefficients are predicted in the Standard Model as function of the Higgs boson mass. We have already illustrated the experimental constraints which exist at least on the coefficients related to the gauge-boson vacuum polarizations. It is therefore a burden of anybody who wants to defend a theory without a Higgs boson to compete with the Standard Model in predicting the coefficients appearing in $\mathcal{L}_{\text{EWCh}}$.

An important consequence is anyhow made clear by this approach to a Higgsless theory. By expanding *e.g.* the third term in (4.24) in powers of the π-fields one finds a first interaction term of the form $\frac{1}{v^2}\pi^2(\partial_\mu \pi)^2$, which grows with the energy as $(E/v)^2$. This shows that a theory without a Higgs boson leads to a saturation of unitarity or, more precisely, to a loss of perturbativity at an energy $E \approx 4\pi v$, unless some extra physics not included in (4.24) comes in before such energy scale. [*Problem 4.6.2: Show explicitly that a one loop correction to the 4 pion amplitude becomes comparable to the tree level term at $E \approx 4\pi v$.*]

Chapter 5
Flavour physics

5.1. The theorems of flavour physics

As we have seen in Lecture 2, the very distinction among the different flavours arises in the Standard Model only after introducing the Yukawa Lagrangian, *i.e.* the sector of the entire Lagrangian which contains the greatest number of free parameters, many more than those appearing in the minimal gauge Lagrangian. As such it may be considered as the least satisfactory among the different sectors of the Standard Model. Nevertheless it is important to realize that it implies a few neat consequences, which we call *theorems* to emphasize their exact nature. Neglecting neutrino masses (an excellent approximation, as we shall see, except when considering neutrino oscillations), they are:

1. There is no flavour transition in the lepton sector.
2. In the quark sector, the only flavour transitions occur in the charged current interaction amplitude depicted in Figure 5.1, $d_i \to W^- u_j$ (or in its hermitian conjugate $u_j \to W^+ d_i$) whose flavour dependence is described by a unitary matrix, V_{ji}.
3. Apart from a possible effect in the strong interactions (see the next lecture), CP violation can take place in some physical process only to the extent that the matrix V just defined is complex.

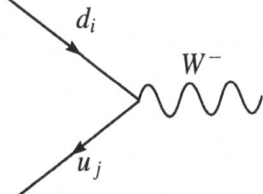

Figure 5.1. The charged current interaction amplitude proportional to V_{ji}.

While we postpone the proof of the third statement to the next lecture, the first two arise as immediate consequences of the Standard Model Lagrangian, as written in equation (2.11). This is manifest for Theorem 1, which expresses individual lepton number conservation, whereas Theorem 2 follows from the redefinition (3.14), needed to go to the physical basis of the left-handed up-type quarks, and the form (3.16) of the couplings to fermions of the gauge bosons. To see this, consider first the interactions of the photon or the Z (or, for that matter, the interactions of the gluons). Since Q and T_3 are diagonal, the redefinition (3.14) does not affect their interactions which remain diagonal also in flavour space. This is not true, on the contrary, for the charged weak interactions, since $T^\pm$ are off diagonal, so that, after the redefinition (3.14), they become explicitly

$$\mathcal{L}_q^{\text{ch-curr}} = \frac{g}{\sqrt{2}} W_\mu^+ \bar{u} V \gamma_\mu d + h.c., \qquad (5.1)$$

where we have identified V_q with V, adopting a standard notation for this matrix, called Cabibbo-Kobayashi-Maskawa matrix.

5.2. Individual lepton number conservation

As already mentioned in Lecture 2 and expressed by Theorem 1, the conservation of the individual lepton numbers implied by the flavour structure of the Standard Model in the massless-neutrino limit is not the least of its successes. A striking confirmation of this is represented by the negative searches for lepton-flavour violating decays of the muon, with the current bounds (CR = Capture Rate)

$$\begin{aligned} \text{BR}(\mu \to e\gamma) &< 1.2\ 10^{-11}, \\ \text{BR}(\mu \to e\bar{e}e) &< 1.0\ 10^{-12}, \\ \text{CR}(\mu \to e \text{ in } Ti) &< 6.1\ 10^{-13}. \end{aligned} \qquad (5.2)$$

On the other hand, since the presence in $\mathcal{L}_{\nu\text{SM}}$ of the right-handed neutrinos introduces, unlike in $\mathcal{L}_{\text{SM}}$, neutrino masses and lepton-flavour violation, as desired to account for neutrino oscillations, it is of interest at least in principle to know the expected size of the branching ratios in (5.2). Focussing our attention to the decay $\mu \to e + \gamma$, one can readily estimate from a diagram like the one of Figure 5.2, a contribution to a flavour-violating effective operator

$$\mathcal{L}_{\mu \to e+\gamma} \approx e \frac{\alpha}{\pi} \frac{v^3}{m_W^4} \left(\lambda^N \min\left(1, \frac{m_W^2}{M^2}\right) (\lambda^N)^+ \lambda^E \right)_{\mu e} (\bar{\mu}\sigma_{\nu\mu}e) F_{\nu\mu}, \qquad (5.3)$$

where λ^N and λ^E are the matrices appearing in equation (2.5) and M is the right-handed-neutrino mass matrix of equation (2.3).

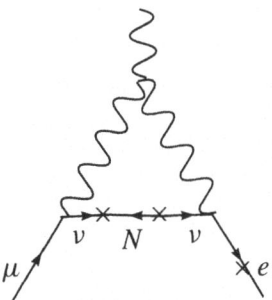

Figure 5.2. A diagram contributing to the $\mu \to e + \gamma$ transition. The crosses denote appropriate mass insertions.

Anticipating from Lecture 7, we have, for the typical size of the elements of the matrix λ^N in terms of the physical neutrino mass m_ν,

$$(\lambda^N)^2 \approx \frac{m_\nu^2}{v^2} \max\left(1, \frac{M}{m_\nu}\right), \qquad (5.4)$$

so that the operator in (5.3) can be rewritten as

$$\mathcal{L}_{\mu \to e+\gamma} \approx A(\bar{\mu}\sigma_{\nu\mu}e)F_{\nu\mu}, \qquad (5.5)$$

with

$$A < e\frac{\alpha}{\pi}\frac{m_\mu}{m_W^2}\frac{m_\nu}{m_W}. \qquad (5.6)$$

The smallness of the observed neutrino masses, below about 1 eV, makes this a totally negligible effect. It is important to realize, however, that the smallness of this effect is not to be traced back to the smallness of the the Yukawa couplings λ_N, which would not be small at all for M large enough, up to $M \approx v^2/m_\nu$ (see equation (5.4)). Rather what suppresses the effect is the exchange in the diagram of Figure 5.2 of the right-handed neutrino(s) of mass M itself (see equation (5.3)). In other words, the decay $\mu \to e + \gamma$ could acquire a significant branching ratio, at the level of the current bound, if the necessary breaking of the lepton flavour symmetry were still be accounted for by λ^N and λ^E, but these same couplings were felt also by some new particle with typical weak scale mass, that could be exchanged in a diagram similar to the one in Figure 5.2. Such particles are present in motivated extensions of the Standard Model.

5.3. About the unitarity of the Cabibbo-Kobaiashi-Maskawa matrix

The quantitative tests of the flavour sector of the Standard Model mainly rest on the verification of the unitarity of the Cabibbo-Kobaiashi-Maskawa matrix, for which a standard self-explanatory notation is

$$V = \begin{pmatrix} V_{ud} & V_{us} & V_{ub} \\ V_{cd} & V_{cs} & V_{cb} \\ V_{td} & V_{ts} & V_{tb} \end{pmatrix}. \qquad (5.7)$$

The unitarity of V implies three relations that only involve the moduli of the matrix elements

$$\Sigma_{i=d,s,b}|V_{ai}|^2 = 1, \qquad a = u, c, t, \qquad (5.8)$$

and three relations that also depend on their phases

$$\Sigma_{i=d,s,b} V_{ai}(V_{bi})^* = 0, \qquad a \neq b = u, c, t. \qquad (5.9)$$

This difference is important in view of CP violation (and Theorem 3). Due to the relatively limited knowledge of the couplings of the top quark, the analogous relations involving a sum over the charge-2/3 quarks are phenomenologically less interesting.

Among the phase-independent relations (5.8), the most stringent test is provided by the $a = u$ case,

$$|V_{ud}|^2 + |V_{us}|^2 + |V_{ub}|^2 = 1, \qquad (5.10)$$

with precision data provided by the measurements of various semileptonic branching ratios, most notably:

- For V_{ud} the superallowed ($J^P = 0^+ \to 0^+$) nuclear transitions $N \to N'e\nu$;
- For V_{us} the semileptonic kaon decays $K \to \pi e\nu$;
- For V_{ub} the semileptonic decays of b-hadrons into final states not containing charm $B \to X_u l\bar{\nu}$.

Whereas the smallness of V_{ub} makes its contribution to (5.10) practically irrelevant, an important uncertainty in the determinations of both V_{ud} and V_{us} arises from the fact that the measurements are not sensitive to the CKM matrix elements alone but rather to the combinations

$$|V_{ud} < f|\bar{u}\gamma_\mu d|i>|, \quad |V_{us} < f|\bar{u}\gamma_\mu s|i>|, \qquad (5.11)$$

where $|i>$ and $|f>$ are the appropriate initial and final hadronic states. Note that the parity selection rule makes it such that it is the vector component of the currents that contribute to the matrix elements in (5.11). In turn these currents, $(\bar{u}\gamma_\mu d)$ and $(\bar{u}\gamma_\mu s)$, would be conserved by the strong interactions in the limit where the global $SU(2)$ or $SU(3)$ symmetries of QCD were exact, implying the precise knowledge, in this limit, of their matrix elements in (5.11). Since both $SU(2)$ and especially $SU(3)$ are only approximate symmetries, this limits the test of equation (5.10), which is however verified to the level of about $1 \div 2\ ppm$.

Turning now to the phase-dependent relations (5.9), each of them, for fixed a and b, can be visualized as a triangle in the complex plane, where every edge, each with an arrow, represents the complex number $V_{ai}(V_{bi})^*$ for the three different $i = d, s, b$. The fact that the three vectors add up to form a closed triangle is the manifestation of the unitarity relation. Note that a change of the phases of the quark fields in equation (5.1) changes also the phase of the matrix elements themselves. As readily seen, this change of phase, which is physically irrelevant, does not affect, however, the shape of the various triangles, but only their orientation relative to a fixed coordinate axis in the complex plane, which is therefore conventional. Given the actual values of the $|V_{ai}|$, out of the three relations (5.9) only one of them,

$$V_{ud}(V_{td})^* + V_{us}(V_{ts})^* + V_{ub}(V_{tb})^* = 0, \qquad (5.12)$$

corresponds to a triangle with edges of comparable size. We shall come back to it in the next lecture.

5.4. Calculable flavour changing neutral current processes

The second flavour Theorem of Section 5.1 not only states that the amplitudes for the charged current transitions in Figure 5.1 are proportional to the elements of a unitary matrix; it also says that these are the only transitions where a change of flavour can take place. In particular, as already said, no change of flavour occurs in the neutral current interaction mediated by the Z-boson exchange. This a neat consequence of the flavour structure of the Standard Model which motivates the notion of *Flavour Changing Neutral Current processes* and gives to them a special interest, both as a test of the Standard Model and as an opportunity to look for new physics not included in the Standard Model itself.

Let us consider any transition existing in the Standard Model with change of flavour and with no W or Z bosons as external states. In order for this transition to take place at all, the exchange of at least a virtual

W is necessary. When reduced to the quark level, we call this transition a Flavour Changing Neutral Current process if it does not proceed by a single virtual W exchange at tree level but requires the virtual W to be exchanged in a loop. As a consequence, in the Standard Model such processes have a reduced rate relative to a normal weak interaction process and their experimental study may uncover the existence of some new interaction. This is especially true for those transitions whose rates are more neatly calculable in the Standard Model. The calculability of any transition involving hadrons is indeed limited by the difficulty of solving QCD in the infrared. For example, the matrix elements among physical hadronic states of a quark effective interaction may be difficult to obtain. Such a problem can however be attacked by computer calculations of QCD on a lattice. A more serious difficulty is when the transition in question is not even reliably calculable at the quark level.

To illustrate this kind of problems, consider the Flavour Changing Neutral Current transition $s\bar{d} \to \bar{s}d$ with a change of two units of strangeness, or $\Delta S = 2$. This transition gives rise to the mixing between the K_0 and the $\bar{K}_0$ (or to the measured mass splitting between the neutral-kaon mass-eigenstates) as well as to the first measured CP violation effect, always in the neutral kaon system (see the next lecture). At the quark level it is approximately induced by the loop diagram of Figure 5.3. The approximation is in the fact that this diagram must be dressed by gluon exchanges among the quark lines with a strong coupling constant g_S that is perturbative only if the typical momentum flowing in the loop is large enough relative to a typical hadronic scale. This becomes therefore the key question to examine.

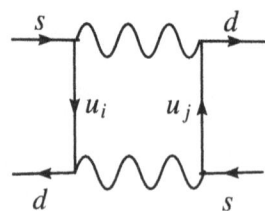

Figure 5.3. A quark diagram contributing to the $s\bar{d} \to \bar{s}d$ transition.

The amplitude corresponding to the diagram of Figure 5.3 with the external momenta set to zero is given by

$$\mathcal{M} = \frac{ig^4}{2} \int \frac{d^4k}{(2\pi)^4} D_{\mu\nu} D_{\sigma\rho} (\bar{d}_L \gamma_\mu S \gamma_\rho s_L)(\bar{d}_L \gamma_\nu S \gamma_\sigma s_L) \qquad (5.13)$$

where
$$D_{\alpha\beta} = \frac{-g_{\alpha\beta} + k_\alpha k_\beta}{k^2 - m_W^2}, \qquad (5.14)$$

$$S = \sum_{i=u,c,t} \frac{\xi_i}{\slashed{k} - m_i}, \qquad \xi_i = V_{is}(V_{id})^* \qquad (5.15)$$

and d_L, s_L are the lower components of the $SU(2)$ quark doublets. Using the unitarity condition $\Sigma_i \xi_{i=u,c,t} = 0$, S can be rewritten as

$$S = \sum_{j=c,t} \xi_j \left(\frac{1}{\slashed{k} - m_j} - \frac{1}{\slashed{k}} \right), \qquad (5.16)$$

where the up quark mass has been neglected. In this way, after reducing the γ-matrix algebra, the overall amplitude can be written in a straightforward way as the sum of three pieces

$$\mathcal{M} = (\xi_t^2 F_{tt} + \xi_c^2 F_{cc} + 2\xi_t \xi_c F_{tc})(\bar{d}_L \gamma_\mu s_L)(\bar{d}_L \gamma_\mu s_L) \qquad (5.17)$$

where

$$F_{ij} = \frac{ig^4}{2} \int \frac{d^4k}{(2\pi)^4} \frac{1 - 2k^2/m_W^2 + (k^2)^2/m_W^4}{k^2(k^2 - m_W^2)^2} \frac{m_i^2}{k^2 - m_i^2} \frac{m_j^2}{k^2 - m_j^2}. \qquad (5.18)$$

This amplitude can be interpreted as an effective Lagrangian, whose matrix element between the K_0 and the $\bar{K}_0$ states gives the mixing matrix M_{12} referred to above. In turn, $2|M_{12}|$ is approximately equal to the mass difference between the physical neutral kaon states, whereas the imaginary part of M_{12} controls the CP violation in the mixing of the same neutral kaon system. Now the values of the CKM matrix elements and of the quark masses are such that the real part dominates M_{12} and is predominantly given by the term proportional to F_{cc} in (5.17). Therefore, since

$$F_{cc} \approx \frac{ig^4}{2m_W^4} \int \frac{d^4k}{(2\pi)^4} \frac{m_c^4}{k^2(k^2 - m_c^2)^2} = \frac{G_F^2 m_c^2}{\pi^2}, \qquad (5.19)$$

the relevant momentum in the loop is between 0 and m_c, hence not such that the gluonic corrections to it are under perturbative control: $2|M_{12}|$ serves only as an estimate of the mass difference between the neutral kaon states, which does not correspond therefore to a calculable Flavour Changing Neutral Current process. On the other hand, as explained in the next lecture, the amplitude physically relevant to CP violation is proportional to the imaginary part of

$$(\xi_u^2)^* M_{12} = (\xi_u^* \xi_t)^2 F_{tt} + (\xi_u^* \xi_c)^2 F_{cc} + 2(\xi_u^* \xi_t)(\xi_u^* \xi_c) F_{tc}. \quad (5.20)$$

Noticing that, again from the unitarity relation, one has $\xi_u^*(\xi_u + \xi_c + \xi_t) = 0$ or

$$\mathcal{I}m(\xi_u^* \xi_t) = -\mathcal{I}m(\xi_u^* \xi_c) \equiv J, \quad (5.21)$$

it is

$$\mathcal{I}m((\xi_u^2)^* M_{12}) = 2J[\mathcal{R}e(\xi_u^* \xi_t)(F_{tt} - F_{tc}) - \mathcal{R}e(\xi_u^* \xi_c)(F_{cc} - F_{tc})]. \quad (5.22)$$

and now, unlike the case for $\mathcal{R}e(M_{12})$, the relevant integration momenta in $(F_{tt} - F_{tc})$ and $(F_{cc} - F_{tc})$ are between m_t and m_c. As a consequence, the gluonic corrections to $\mathcal{I}m((\xi_u^2)^* M_{12})$ are under control: CP violation in the mixing of the neutral kaon system corresponds to a calculable Flavour Changing Neutral Current process and enters among the observables used for a quantitative test of the flavour sector of the Standard Model.

5.5. Summary of calculable FCNC processes

Totally analogous considerations can be made for the $\Delta B = 2$ Flavour Changing Neutral Current transitions $b\bar{d} \to \bar{b}d$ and $b\bar{s} \to \bar{b}s$, which again manifest themselves in a mass difference and in CP violation in the neutral B-meson states. The main difference, on the other hand, is in the CKM matrix elements relevant to the b-quark rather than the s-quark: the contribution of the virtual top quark in the diagram of Figure 5.3, with s replaced by b, becomes relatively more important. Consequently, both the mass difference and the CP violating effects become calculable.

In Table 5.1 one has the list of the calculable FCNC processes that have either already been measured or are expected to be measured, with the current experimental uncertainties and an estimate of the present theoretical uncertainties only due to the sources described in the previous Section, *i.e.* with CKM matrix elements assumed perfectly known. A common feature of all the FCNC processes listed in Table 5.1, with the partial exception of CP violation in the kaon system, as discussed above, is the dominance of the virtual top exchange in the loop-induced FCNC transition. All the measured FCNC processes in this table are at present

Table 5.1. List of calculable Flavour Changing Neutral Current processes. The blank boxes are because of absence of data so far.

Observable	elementary process	exp. error	theor. error
ϵ_K	$\bar{s}d \to \bar{d}s$	1%	$10 \div 15\%$
$K^+ \to \pi^+ \bar{\nu}\nu$	$s \to d\,\bar{\nu}\nu$	70%	3%
$K^0 \to \pi^0 \bar{\nu}\nu$	$s \to d\,\bar{\nu}\nu$		1%
Δm_{Bd}	$\bar{b}d \to \bar{d}b$	1%	25%
$A_{CP}(B_d \to \Psi K_S)$	$\bar{b}d \to \bar{d}b$	5%	< 1%
$B_d \to X_s + \gamma$	$b \to s + \gamma$	10%	$5 \div 10\%$
$B_d \to X_s + \bar{l}l$	$b \to s + \bar{l}l$	50%	$5 \div 10\%$
$B_d \to X_d + \gamma$	$b \to d + \gamma$		$10 \div 15\%$
$B_d \to \bar{l}l$	$b\bar{d} \to \bar{l}l$		10%
$B_d \to X_d + \bar{l}l$	$b \to d + \bar{l}l$		$10 \div 15\%$
Δm_{Bs}	$\bar{b}s \to \bar{s}b$	< 1%	25%
$A_{CP}(B_s \to \Psi\phi)$	$\bar{b}s \to \bar{s}b$		1%
$B_s \to \bar{l}l$	$b\bar{s} \to \bar{l}l$		10%

consistent with their description in the Standard Model and with the direct determinations of the CKM matrix elements from tree-level flavour-changing transitions.

Chapter 6
CP violation

6.1. The source(s) of CP violation in the Lagrangian of the Standard Model

Let us start by proving the third flavour theorem of the previous lecture. To this purpose it is best to work in the unitary gauge for the vector bosons and in the physical basis for the fermion fields. The transformation laws under CP of the fermion bylinears are given in Appendix E together with the general transformations of the vector bosons. Specialized to the electroweak bosons they are

$$A_\mu(\mathbf{x}, t) \rightarrow -A_{\tilde{\mu}}(-\mathbf{x}, t), \quad Z_\mu(\mathbf{x}, t) \rightarrow -Z_{\tilde{\mu}}(-\mathbf{x}, t), \tag{6.1}$$

and

$$W_\mu^+(\mathbf{x}, t) \rightarrow -W_{\tilde{\mu}}^-(-\mathbf{x}, t). \tag{6.2}$$

Since under CP, for any pair of left handed spinors

$$\bar{\psi}\gamma_\mu \chi \rightarrow -\bar{\chi}\gamma_{\tilde{\mu}}\psi(-\mathbf{x}, t) \tag{6.3}$$

the electromagnetic and the neutral current interactions, being diagonal, are manifestly invariant under CP, after space-time integration. On the contrary, the charged current interaction (5.1) becomes

$$\mathcal{L}_q^{\text{ch-curr}} \rightarrow \frac{g}{\sqrt{2}} W_\mu^+ \bar{u} V^* \gamma_\mu d(-\mathbf{x}, t) + h.c., \tag{6.4}$$

which is invariant only if V is real, as anticipated. On the other hand, from

$$\bar{\psi}\chi \rightarrow \bar{\chi}\psi, \tag{6.5}$$

the Higgs boson interactions with the fermions, equation (3.22), with the vectors, equation (3.21), as well as the Higgs self interactions, equation (3.19), are all invariant under CP, provided

$$h(\mathbf{x}, t) \rightarrow h(-\mathbf{x}, t). \tag{6.6}$$

This proves the theorem about CP violation in the Standard Model. Notice however that, from the proof given above, a complex V is only a necessary condition for the existence of some CP violating observable. For it to be sufficient, it must not be possible to make V real by a redefinition of the unphysical quark phases. It is in fact useful to discuss this condition in the case of an arbitrary number n of different generations. Remember to this purpose that u and d in equation (6.4) stand for column vectors with as many elements as the number of generations. In general an arbitrary $n \times n$ unitary matrix depends upon $n^2 = (1/2)n(n-1) + (1/2)n(n+1)$ real numbers: $(1/2)n(n-1)$ are the elements of an ortogonal $n \times n$ matrix, or the number of rotation angles, whereas the remaining $(1/2)n(n+1)$ parameters are phases. Since in the charged current in (6.4) there are n up quarks and n down quarks, it would seem that $2n$ of these $(1/2)n(n+1)$ phases can be redefined away. This is not correct, however, since one of the $2n$ quark phases, the one that corresponds to the conserved baryon-number symmetry cannot change the current at all. Hence the number of phases which cannot be eliminated is $(1/2)n(n+1) - (2n-1)$. This is a remarkable result, since it says that the number of generations n must be equal to three to have at least one phase available to describe a physical CP violation. This observation was made before the experimental discovery of the third generation.

A final qualification is necessary for the theorem about CP violation in the Standard Model, other than recalling that we are neglecting for the time being any effect due to neutrino masses. As shown in Appendix E, the field strenghts associated to any generator t^a transform under CP as

$$F_{\mu\nu} \equiv F_{\mu\nu}^a t^a \to -F_{\tilde{\mu}\tilde{\nu}}^T(-\mathbf{x}, t). \tag{6.7}$$

Therefore, unlike the gauge kinetic term $F_{\mu\nu}^a F_{\mu\nu}^a$ which is invariant after space-time integration, $\epsilon_{\mu\nu\rho\sigma} F_{\mu\nu}^a F_{\rho\sigma}^a$ is odd under CP. We have not included such term in the basic Lagrangian (2.7) for any of the gauge-group factors since it can be written as a total space-time derivative. It is in fact

$$\epsilon_{\mu\nu\rho\sigma} F_{\mu\nu}^a F_{\rho\sigma}^a = \frac{1}{2}\epsilon_{\mu\nu\rho\sigma} Tr[F_{\mu\nu} F_{\rho\sigma}] = \partial_\rho J_\rho \tag{6.8}$$

where

$$J_\rho = 2\epsilon_{\mu\nu\rho\sigma} Tr[A_\sigma \partial_\mu A_\nu + \frac{2i}{3} A_\mu A_\sigma A_\nu] \tag{6.9}$$

and $A_\mu = A_\mu^a t^a$. However for a non Abelian gauge group even a total derivative like (6.8) contributes to the action through boundary terms. With respect to possible CP violating effects in the Standard Model, this is especially important for the strong $SU(3)$ gauge group. An addition to the total Lagrangian of a dimension-4 term

$$\Delta \mathcal{L}_{\text{strong CP}} = \theta_{\text{QCD}} \epsilon_{\mu\nu\rho\sigma} G^a_{\mu\nu} G^a_{\rho\sigma} \tag{6.10}$$

where G^a_μ is the gluon field, would lead to observable CP violating effects (see next section) unless the dimensionless parameter θ_{QCD} is very small, below about 10^{-9}. Why such parameter is so small goes under the name of *strong CP problem*.

6.2. Electric dipole moments

Any conceivably measurable effect of CP violation in the Standard Model only involves the light degrees of freedom, *i.e.* the gluons, the photon and all the quarks, except the top. To describe these effects it is therefore useful to consider the effective Lagrangian relevant at a scale well below the W and the Z masses. Given the unbroken group of gauge invariance, $SU(3) \times U(1)_{em}$, one can write down the form of this Lagrangian much in the same way as it was done in the first Lectures for the full Standard Model. The difference is that we have to be prepared to include also terms of dimension higher than 4 (see Appendix D). In fact it follows from the discussion made for the Standard Model that one cannot write down any CP violating term in the effective Lagrangian of dimension less then or equal to 4 with the only possible exception of (6.10). To reach this conclusion it is essential, as usual, that we allow for possible field redefinitions, which are physically irrelevant. It is also of importance to know that a strong CP term like (6.10) is not generated at a significant level, *i.e.* with θ_{QCD} above or even close to 10^{-9}, if it is not there in the original Lagrangian, after integration of the heavy degrees of freedom up to a cut-off as large as the Planck mass.

We are therefore led to consider terms in the effective Lagrangian of dimension greater than 4. At dimension 5, an operator of key interest is

$$\mathcal{L}_{\text{DM}} = e\mu_q e^{i\delta}(\bar{q}_L i\sigma_{\nu\mu} q_R) F_{\nu\mu} + h.c., \tag{6.11}$$

where e is the electron charge, q is either a up or a down quark, μ is a real parameter of dimension of $mass^{-1}$, $F_{\nu\mu}$ is the electromagnetic field strenght and δ is a phase with the left and right components of the quark field, q_L, q_R, defined in such a way that the corresponding mass term in the effective Lagrangian, $m_q \bar{q}_L q_R + h.c.$, is real. Since under CP (see Appendix E)

$$\bar{q}_L i\sigma_{\nu\mu} q_R \to -\bar{q}_R i\sigma_{\tilde{\mu}\tilde{\nu}} q_L, \tag{6.12}$$

$\mathcal{L}_{\text{DM}}$ violates CP if δ is different from zero.

It is a simple matter to obtain from (6.11) the corresponding scattering amplitude from a static electromagnetic field in the non relativistic limit for the quark spinors. If ξ and ξ' are the non relativistic spin wave functions of the initial and final quark states, one gets

$$i\mathcal{M}_{\text{DM}} = ie\mu_q \xi'^+ \left(\cos\delta \frac{\sigma}{2} \cdot \mathbf{B} + \sin\delta \frac{\sigma}{2} \cdot \mathbf{E} \right) \xi \qquad (6.13)$$

where $\mathbf{B}$ and $\mathbf{E}$ are the static magnetic and electric fields in Fourier space. The first term in the right-hand-side of (6.13) is a magnetic moment interaction, present also in the non relativistic limit of the standard dimension-4 interaction term, whereas the second term, only present for $\delta \neq 0$, is a CP violating interaction with the electric field through an electric dipole moment

$$\mathbf{d}_E = e \sin\delta \mu_q \mathbf{S}, \qquad (6.14)$$

where $\mathbf{S}$ is the spin of the quark. [*Problem 6.2.1: Prove equation (6.13).*] An electric dipole moment for the quarks u or d gives rise to an electric dipole moment for the neutron of roughly similar order. Such an intrinsic dipole moment for the neutron has been searched for with negative results so far, so that currently

$$\mathbf{d}_E(\text{neutron}) \leq 6 \cdot 10^{-26} e \cdot cm. \qquad (6.15)$$

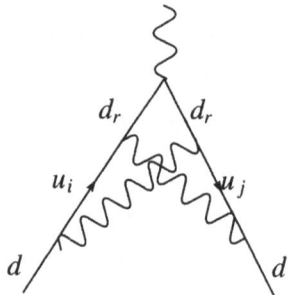

Figure 6.1. A 2 loop diagram contributing to the electric dipole moment of the d-quark.

This bound is well above the expected theoretical value of the neutron electric dipole moment, which can be estimated as follows. No $\mathcal{L}_{\text{DM}}$ with non vanishing δ is generated from a one loop W-exchange diagram, which is proportional to $|V_{qi}|^2$. To get a contribution to the electric dipole moment one has to go to two loop order, as from the diagram of Figure 6.1. We are then led to the estimate

$$|\mathbf{d}_E(q)| \approx e \left(\frac{\alpha}{4\pi} \right)^2 |J| \frac{m_q}{m_W^2}, \qquad (6.16)$$

where J is defined in equation (5.21) and the mass of the quark q appears in the numerator due to the necessary breaking of the associated chirality. $\mathbf{d}_E(q)$ is proportional to J since any quadrilinear product of V_{ij} which is invariant under redefinition of the quark phases is either real or it has an imaginary part equal to $\pm J$. This can be proved, as the special case of equation (5.21), by repeated use of the ortogonality relations between different rows and columns of the CKM matrix, like equation (5.9). For this reason J is called Jarlskog invariant. [*Problem 6.2.2: Show this property of J.*] From the current values of the measured CKM parameters, $J \approx 3 \cdot 10^{-5}$. For a quark mass of about 10 MeV, we get therefore from equation (6.16) the order of magnitude estimate, $\mathbf{d}_E(q) \approx 10^{-30} e \cdot cm$, which is indeed far from the current experimental limit. The situation is quite different, however, in several motivated extensions of the Standard Model which contain new particles with typical weak scale mass, much as in the case of the $\mu \to e + \gamma$ transition, as commented upon in the previous Lecture, or of the same electric dipole moment of the electron $\mathbf{d}_E(e)$. From current experiments, $|\mathbf{d}_E(e)| \lesssim 10^{-27} e \cdot cm$, while it is completely negligible in the Standard Model.

6.3. CP violation in effective 4-fermion interactions

Where CP violation has been experimentally observed is from effects due to 4-fermion interactions, of dimension 6, both with change of strangeness and of beauty. For concreteness we discuss in the following the case of strangeness changing transitions. The relevant effective Lagrangian is

$$\mathcal{L}_{\text{eff}}^{(\Delta S \neq 0)} = \mathcal{L}_{\text{eff}}^{(\Delta S=2)} + \mathcal{L}_{\text{eff}}^{(\Delta S=1)} + \mathcal{L}_{\text{eff}}^{(\Delta S=1;\text{semilept.})} + h.c, \quad (6.17)$$

where, somewhat schematically,

$$\mathcal{L}_{\text{eff}}^{(\Delta S=2)} = A(\bar{s}_L \gamma_\mu d_L)(\bar{s}_L \gamma_\mu d_L), \quad (6.18)$$

$$\mathcal{L}_{\text{eff}}^{(\Delta S=1)} = \Sigma_q B_q (\bar{s}_L \gamma_\mu d_L)(\bar{q} \gamma_\mu q), \quad (6.19)$$

$$\mathcal{L}_{\text{eff}}^{(\Delta S=1;\text{semilept.})} = \Sigma_l C_l (\bar{s}_L \gamma_\mu d_L)(\bar{l} \gamma_\mu l). \quad (6.20)$$

In equation (6.19) the sum extends over the u and the d quarks of both elicities, as arising in particular from the gluon exchange diagrams of Figure 6.2, whereas in (6.20) the sum extends over the leptons, either charged or neutral.

The coefficients A, B_q, C_l, of dimension of $mass^{-2}$, are in general complex and it is their relative phase which amounts to a CP violation.

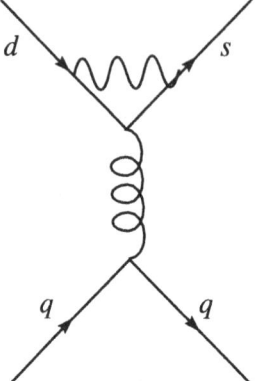

Figure 6.2. A so called *penguin* diagram contributing to the $\Delta S = 1$ transition in equation (6.19).

We have discussed in the previous Lecture how to calculate the imaginary part of A (with a suitable definition of the quark phases). In particular a relative phase which cannot be redefined away by changing the phases of the quark fields can occur:

1. Between A and any of the B_q;
2. Among the B_q themselves;
3. Between A and any of the C_l.

Case 1 is the source of CP violation first observed in Kaon physics as well as the one seen in more decay channels, all involving the neutral Kaons. The amplitude A was discussed in Section 5.4, whereas the dominant among the B_q is due to the tree level W-exchange with $q = u_L$, proportional to $\xi_u = V_{us}(V_{ud})^*$. This explains why the physical CP violating effect is obtained by looking at the the imaginary part of the amplitude M multiplied by $(\xi_u^*)^2$, as it was done in Section 5.4, since this is the re-phasing invariant combination. In the neutral Kaon system, the eigenstates of the propagation Hamiltonian

$$i\frac{d}{dt}|K_{L,S}>= \left(m_{L,S} - \frac{i}{2}\gamma_{L,S}\right)|K_{L,S}> \qquad (6.21)$$

are superpositions of the states with definite strangeness

$$|K_{L,S}>= \frac{1}{\sqrt{|p|^2 + |q|^2}}(p|K^0> \pm q|\bar{K}^0>). \qquad (6.22)$$

Since under CP $|K^0> \to e^{i\delta}|\bar{K}^0>$ and $|\bar{K}^0> \to e^{-i\delta}|K^0>$, the eigenstates of the propagation Hamiltonian are not CP eigenstates if $|p/q| \neq 1$, in which case CP is violated. By relating A and B_u to the effective propagation Hamiltonian of the neutral Kaons, one shows in fact that

$$\left|\frac{p}{q}\right| \propto \mathcal{I}m((\xi_u^2)^* M_{12}), \tag{6.23}$$

in turn related to the parameter ϵ_K listed in Table 5.1. This so called *indirect* CP violation controls the measured parameters

$$\delta_K = \frac{\Gamma(K_L \to \pi^- l^+ \nu) - \Gamma(K_L \to \pi^+ l^- \bar{\nu})}{\Gamma(K_L \to \pi^- l^+ \nu) + \Gamma(K_L \to \pi^+ l^- \bar{\nu})} \tag{6.24}$$

$$= (3.27 \pm 0.12) 10^{-3},$$

$$|\eta_{00}| \equiv \left|\frac{A(K_L \to \pi^0 \pi^0)}{A(K_S \to \pi^0 \pi^0)}\right| = (2.276 \pm 0.014) 10^{-3},$$

$$|\eta_{+-}| \equiv \left|\frac{A(K_L \to \pi^+ \pi^-)}{A(K_S \to \pi^+ \pi^-)}\right| = (2.288 \pm 0.014) 10^{-3}. \tag{6.25}$$

which all violate CP. In the case of (6.24), the final states $\pi^- l^+ \nu$ and $\pi^+ l^- \bar{\nu}$ integrated over momenta and summed over spins are the CP conjugate of each other and, in the case of (6.25), the 2π system with zero angular momentum has definite CP, whereas K_L and K_S, if they were CP eigenstates, would have opposite CP. [*Problem 6.3.1: Calculate δ_K in terms of $|p/q|$.*]

The CP violation observed in (6.24) and (6.25) is called indirect because it arises from the mixing of states of opposite CP. As opposed to this case, the CP violation originated by a physical phase among the $\Delta S = 1$ amplitudes B_q (case 2 above) is called *direct*. Since it does not require mixing of the initial states, it can in principle take place both in neutral and in charged Kaon decays. Direct CP violation requires on the other hand at least two physically different decay amplitudes capable of interfering among each other. The only observed direct CP violation in the Kaon system is the deviation from unity of the ratio $|\eta_{00}/\eta_{+-}|$, proportional to the parameter

$$\epsilon' \equiv \frac{1}{2\sqrt{2}}\left(\frac{A_2}{A_0} - \frac{\bar{A}_2}{\bar{A}_0}\right) \tag{6.26}$$

where

$$A_I \equiv \sqrt{\frac{2}{3}} A(K^0 \to 2\pi, I), \quad \bar{A}_I \equiv \sqrt{\frac{2}{3}} A(\bar{K}^0 \to 2\pi, I) \tag{6.27}$$

and $I = 0, 2$ are the possible isospin states of the two pion system. From Unitarity and CPT invariance, it can be shown that ϵ' deviates from zero to the extent that $\mathcal{A}_2$ and $\mathcal{A}_0$ receive different phases from a physical phase in the B_q's. Due to the large number of operators (or B_q's) that contribute to the $\mathcal{A}_I$ and to the near cancellations that take place among them, it proves hard, however, to go beyond an order of magnitude prediction of ϵ' in the Standard Model. While this prediction agrees with the experimental result

$$1 - |\eta_{00}/\eta_{+-}| = (5.01 \pm 0.78)10^{-3}, \qquad (6.28)$$

its uncertainty does not allow the inclusion of direct CP violation in Table 5.1.

From a theoretical point of view, similar considerations hold for CP violation in the B-system due to 4-fermion interactions with change of beauty (and possibly strangeness as well). Physically the main differences between the B and the K systems are in the much larger number of decay modes of the former and, not unrelated, in the relative widths of the two neutral eigenstates of the weak Hamiltonian

$$\Delta\Gamma|_K \equiv \frac{\gamma_S - \gamma_L}{\gamma_S + \gamma_L} \approx 1 \gg \Delta\Gamma|_B, \qquad (6.29)$$

either for the B_d or the B_s systems. Some entries related to CP violation in the B-system do in fact appear in Table 5.1 together with several other calculable FCNC effects. [*Problem 6.3.2: Estimate $\Delta\Gamma|_B$ for the B_d or the B_s systems. Question 6.3.1: Why no entry appears in Table 5.1 for any charmed meson?*]

Chapter 7
Basics of neutrino physics

7.1. The three options for neutrino masses in the Standard Model

As shown in Lecture 2, the natural way to describe neutrino masses in the Standard Model consists in including three right handed neutrinos N_i among the fermionic degrees of freedom and in adding to the Standard Model Lagrangian the extra renormalizable terms involving them, so that

$$\mathcal{L}_{\nu\text{SM}} = \mathcal{L}_{\text{SM}} + \bar{N}_i\, \slashed{\partial} N_i - \left(\phi L_i \lambda_{ij}^N N_j + \frac{1}{2} N_i M_{ij} N_j + \text{h.c.} \right). \quad (7.1)$$

When ϕ gets a vacuum expectation value, the neutrino mass terms are a 6×6 matrix of the form

$$\mathcal{L}^{(\nu-mass)} = -\frac{1}{2} (\nu^T \ N^T) \begin{pmatrix} 0 & \lambda^N v \\ (\lambda^N)^T v & M \end{pmatrix} \begin{pmatrix} \nu \\ N \end{pmatrix}. \quad (7.2)$$

There are two limiting cases of interest for this mass matrix (see Appendix F for the definition of Dirac and Majorana neutrinos):

- *Three Dirac neutrinos*

This arises when the mass term of the right handed neutrinos in (7.1), $-\frac{1}{2} N^T M N$ is set to zero, thus recovering lepton number conservation. From the point of view of the Yukawa couplings, the leptons become fully analogous to the quarks: lepton number corresponds to baryon number and every fermion in the Standard Model is described by a Dirac spinor.

- *Three light Majorana neutrinos*

The general neutrino mass matrix (7.2) has 6 different eigenvalues, 2 by 2 equal to each other in the previous case of three Dirac neutrinos. If, however, the elements of M are much larger than any element in $\lambda^N v$, the full mass matrix approximately factorizes in two pieces, as readily seen by treating $vv_i\lambda^N_{ij}N_j$ as a perturbation relative to $N_i M_{ij} N_j$, so that

$$\mathcal{L}_{\nu mass} \approx \frac{v^2}{2} v^T \lambda^N \frac{1}{M}(\lambda^N)^T v - \frac{1}{2} N^T MN + \text{h.c.} \quad (7.3)$$

In this way one ends up with three light neutrinos, the eigenvectors of $\lambda^N \frac{1}{M}(\lambda^N)^T$, mostly made of the weakly interacting v_i, and three heavy approximately decoupled right handed neutrinos. Lepton number in this case is maximally violated, even though, for particular forms of M with some degenerate eigenvalues, a combination of individual lepton numbers may still survive.

- *More than three light neutrinos*

It is clear that the two alternatives just described do not exhaust all the possibilities. It is conceivable that there be more than three light and distinct eigenvalues of the full neutrino matrix, in fact up to six. The current phenomenology of neutrino oscillations constrains this possibility but does not exclude it.

Although these three alternatives are all logically viable, the second one is more frequently considered since one prefers to attribute the smallness of the neutrino masses to a large value of the elements of the matrix M rather then to the smallness of the dimensionless couplings in λ^N. With reference to this case, it is in fact

$$m_\nu \approx 0.1\,\text{eV} \left(\frac{\lambda v}{\text{GeV}}\right)^2 \frac{10^{10}\text{GeV}}{M} \quad (7.4)$$

where 0.1 eV is a plausible value for a neutrino mass, as commented upon in the next Lecture, and 1 GeV is in the (actually broad) ballpark of the charged fermion masses.

7.2. The physical parameters

As discussed in Lecture 2, there is a redundancy of parameters in the Yukawa sector which is useful to reduce away. In the case of three Dirac neutrinos this has already been done (see equations (2.11, 2.12)) in full analogy with the quark case. Other than the six masses, three for the charged leptons and three for the neutrinos, $m_i, i = 1, 2, 3$, there is a unitary matrix, V_l, which in the physical basis for the charged leptons, l_i, and the neutrinos, of left-handed components v_i^{phys}, appears in the charged current interaction Lagrangian

$$\mathcal{L}_l^{\text{ch-curr}} = \frac{g}{\sqrt{2}} W_\mu^+ \bar{v}^{\text{phys}} V_l \gamma_\mu l + \text{h.c.} \quad (7.5)$$

We shall frequently drop the superscript *phys* without risks of ambiguity. As we called V the mixing matrix V_q in the quark case, we shall call U^T the matrix V_l to keep the convention most frequently used in the literature. The analogy between V and U is clearly also complete as far as the counting of their physical parameters goes: three angles and one phase.

The situation in the case of three light Majorana neutrinos is slightly different. To find the neutrino mass eigenstates (the *physical* neutrinos), the mass term that has to be diagonalized is the first one in equation (7.3). This is achieved by the redefinition $v_i \to (U^*)_{ij} v_j^{\text{phys}}$, where U is again the matrix that will appear in the charged current interactions. The only difference that arises relative to the Dirac-neutrino case is in the physical phases that enter in U. To understand this difference, we have to go back to the previous lecture where we have discussed the quark case, or by complete analogy the Dirac-neutrino case. There, by shifting the otherwise unphysical phases of the $u(v)$ and $d(l)$ fields, we could subtract away from the $(1/2)n(n+1)$ phases of an arbitrary $n \times n$ unitary matrix the $(2n-1)$ phases that affect the charged current. This is not possible anymore in the Majorana-neutrino case, since the phases of the neutrino fields are physical, due to the form of the Majorana mass matrix $v^T M v$. We can only subtract away the phases of the n charged leptons, thus remaining with $(1/2)n(n-1)$ physical phases, i.e three phases in the three generation case.

If there are more than three light neutrinos, the number of physical parameters obviously increases. A possible choice corresponds to ignore first the Majorana mass matrix M for the right handed neutrinos and go to the same basis as in the case of pure Dirac neutrinos. On top of this situation one has to include the effects of M, whose matrix elements are all in principle physical.

The production or the detection of a neutrino takes place most of the time by a charged current interaction (7.5) in association with the emission or the absorption of a charged lepton $l_\alpha = e, \mu, \tau$

$$\tilde{v}_\alpha \to W^+ + l_\alpha, \qquad l_\alpha \to W^- + \tilde{v}_\alpha, \qquad (7.6)$$

where it is important to realize that $\tilde{v}_\alpha$ is not a physical or a mass eigenstate but rather the superposition of them

$$\tilde{v}_\alpha \equiv U^*_{\alpha i} v_i^{\text{phys}} \qquad (7.7)$$

which appears in the charged current interaction in association with l_α. Here again the tilde on the neutrino will most of the time be omitted.

7.3. Neutrino mass measurements from the β-decay spectrum

The historical way to search for neutrino masses, already proposed by Fermi in his original work mentioned in the first lecture, consists in searching for a distortion of the electron spectrum, dN_e/dE_e, in the β-decay of a nucleus (*i.e.* $d \to ue\bar{\nu}_e$ at the quark level or $n \to pe\bar{\nu}_e$ at the nucleon level).

The most sensitive case is tritium decay

$$^3\text{H} \to {}^3\text{He}\, e\, \bar{\nu}_e, \qquad (Q = m(^3\text{H}) - m(^3\text{He}) = 18.6 \text{ keV}), \qquad (7.8)$$

where $E_e \approx Q - E_\nu$ is required by energy conservation. The most important effect of the neutrino mass m_ν is at the endpoint, $E_e \leq Q - m_\nu$, of the electron spectrum, which is essentially determined by the neutrino phase space factor, proportional to $E_\nu p_\nu$. Therefore, close to the endpoint,

$$\frac{dN_e}{dE_e} \propto (Q - E_e)\sqrt{(Q - E_e)^2 - m_\nu^2}, \qquad (7.9)$$

which means that at the endpoint the spectrum has infinite derivative in the massive neutrino case, instead of a vanishing derivative for a massless neutrino. This is the feature that should be seen by the experiment looking for a neutrino mass with enough energy resolution.

In the case of several neutrino mass eigenstates, equation (7.9) trivially generalizes to

$$\frac{dN_e}{dE_e} \propto \sum_i |U_{ei}|^2 (Q - E_e)\sqrt{(Q - E_e)^2 - m_i^2}, \qquad (7.10)$$

the sum being extended in principle over all the physical neutrinos. In the three neutrino case, either Dirac or Majorana, the only one of practical relevance, the energy resolution of current or foreseen β-decay experiments is unlikely to resolve the difference between neutrino masses, limited by the oscillation experiments (see the next lecture). In this case, it is useful to approximate (7.10) with (7.9) and express the experimental bound in term of a single effective parameter $m_\nu^2 \equiv \Sigma_i |U_{ei}|^2 m_i^2$. At the time of writing these lectures, the upper limit set by β-decay experiments on m_ν is between 1 and 2 eVs.

7.4. Neutrino-less double-β decay

In some cases the β-decay of a nucleus (A, Z) to (A, Z+1) is kinematically forbidden, whereas the double-β decay to (A, Z+2) is allowed. In the Standard Model this is a second order weak interaction process, *i.e.* occurring at the nucleon level through two simultaneous $n \to pe\bar{\nu}_e$ decays. This is the case, *e.g.*, of $^{76}_{32}$Ge that cannot decay to the heavier $^{76}_{33}$As but decays to the lighter $^{76}_{34}$Se with the emission of two electrons and two antineutrinos. The Q-value of this reaction is about 2 MeV and the corresponding life-time is approximately 10^{21} yr.

The great interest of double-β decay is that it can occur also without the emission of the antineutrinos, *i.e.* with violation of lepton number by two units, if neutrinos have Majorana masses. The corresponding Feynman diagram is shown in Figure 7.1 together with the diagram for the normal $2\nu2\beta$ process. The two processes are neatly distinguishable by the spectrum of the two electrons, of definite energy $E = Q/2$ only in the neutrino-less case. The $0\nu2\beta$ decay is apparently the only known process that may lead to a positive evidence for neutrino masses of specific Majorana type.

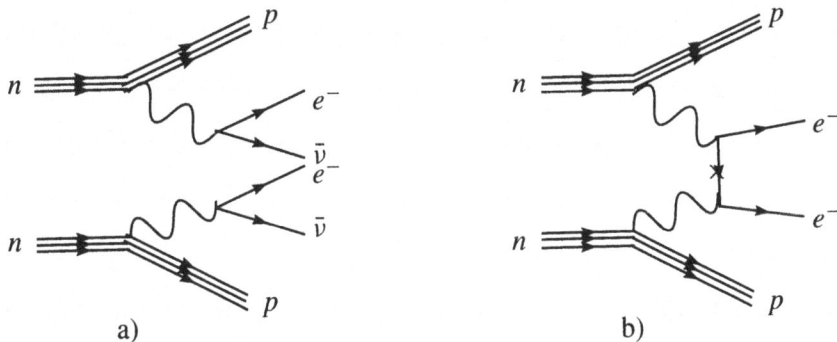

Figure 7.1. Double-β decay amplitudes with 2 neutrinos (a) and without neutrinos (b). The cross in (b) denotes a Majorana mass insertion for the intermediate neutrino.

In the limit in which all the neutrino masses are negligible with respect to Q, the amplitude for Figure 7.1b is linear in the neutrino masses and is therefore actually proportional to the combination

$$m_{ee} \equiv \sum_i U_{ei}^2 m_i. \tag{7.11}$$

In turn the rate takes the form

$$\Gamma_{0\nu2\beta} = |m_{ee}|^2 |\mathcal{M}|^2 \Phi \tag{7.12}$$

where Φ is a known phase space factor and $\mathcal{M}$ is the nuclear $0\nu2\beta$ matrix element, unfortunately plagued by considerable theoretical uncertainties. [*Problem 7.4.1: Calculate* Φ *and find the explicit form of* $\mathcal{M}$.]

The background from natural and artificial sources is a great difficulty for this type of experiments. At this moment it is safe to quote an upper limit of about 1 eV on m_{ee}. Assuming $m_{ee} = 1$ eV, the lifetime for $2\nu 2\beta$ decay of the most promising isotopes is in the range of 10^{24} ys. Anticipating the information gathered on U_{ei} and m_i from oscillation experiments (see next chapter), m_{ee} ranges typically from a fraction of an electronVolt down to a few MeV, although it is strictly speaking not bounded from below.

Chapter 8
Neutrino oscillations

8.1. Neutrino oscillations in vacuum

Neutrino oscillations are a single-particle quantum-mechanical effect of great interest, since to date they represent the only source of information on neutrino masses. They consist in the following phenomenon. A neutrino wave-packet is produced at a source, located at approximately $x = 0$, in association with a charged lepton l_α. The wave packet evolves until it reaches a detector at approximately $x = L$, where a charged lepton l_β, generally different from l_α, is detected. We want to know the probability for this to happen.

To this end we describe the neutrino wave-packet at an intermediate point x as

$$|\nu(x)\rangle = \int dE f(E) \sum_i U^*_{\alpha i} e^{(ip_i x - iEt)} |\nu_i^{phys}\rangle, \qquad (8.1)$$

where $p_i(E) = \sqrt{E^2 - m_i^2}$ is the momentum of the neutrino of energy E and mass m_i, so that at $x = 0$ it is indeed $|\nu(0)\rangle = |\nu_\alpha\rangle$ as defined in (7.7). The amplitude for finding at x a neutrino of energy E in association with a charged lepton l_β is therefore

$$\mathcal{A}_{\beta\alpha}(x) = f(E) \sum_i (U_{\alpha i})^* U_{\beta i} e^{(ip_i x - iEt)}. \qquad (8.2)$$

Now the key assumption which greatly simplifies all the relevant formulae is that, when considering the probability for the oscillation to occur, it is legitimate to neglect all interferences between amplitudes of different energy, so that, for the probability we are interested in,

$$P_{\beta\alpha}(L) = \int dE |f(E)|^2 \left| \sum_i (U_{\alpha i})^* U_{\beta i} e^{ip_i L} \right|^2. \qquad (8.3)$$

This simplifying assumption is valid in all realistic oscillation experiments, due to the properties of the source and/or of the detector. The neutrino wave is fully described by its energy spectrum only, $dN/dE = |f(E)|^2$. Having made clear that we deal with a neutrino wave-packet, so that we avoid running in all sorts of paradoxes, in the following we adopt the standard convention of leaving understood the flux factor $dE|f(E)|^2$.

Taking into account that generally $E \gg m_i$, the oscillation probability simplifies to

$$P_{\beta\alpha}(L) = \left|\sum_i (U_{\alpha i})^* U_{\beta i} e^{2i\phi_i}\right|^2, \quad \phi_i = -\frac{m_i^2 L}{4E} \tag{8.4}$$

or, using $e^{2i\phi} = 1 - 2\sin^2\phi + i\sin 2\phi$,

$$P_{\beta\alpha}(L) = \sum_{ij} J_{ij}^{\beta\alpha} (1 - 2\sin^2\phi_{ij} + i\sin 2\phi_{ij}), \tag{8.5}$$

where

$$\begin{aligned}\phi_{ij} &\equiv \phi_i - \phi_j = \frac{\Delta m_{ij}^2 L}{4E}, \\ \Delta m_{ij}^2 &= m_i^2 - m_j^2, \\ J_{ij}^{\beta\alpha} &= (U_{\alpha i})^* U_{\beta i} (U_{\beta j})^* U_{\alpha j}.\end{aligned} \tag{8.6}$$

Furthermore, using the unitarity relations $\Sigma_i (U_{\alpha i})^* U_{\beta i} = \alpha_{\beta\alpha}$ one gets

$$\begin{aligned}P(\nu_\alpha \to \nu_\beta) = \alpha_{\beta\alpha} &- \sum_{i<j} 4Re(J_{ij}^{\beta\alpha}) \sin^2 \phi_{ij} \\ &- \sum_{i<j} 2Im(J_{ij}^{\beta\alpha}) \sin 2\phi_{ij},\end{aligned} \tag{8.7}$$

where we have made explicit that we deal with neutrinos. The corresponding formulae for antineutrinos are obtained by exchanging $U \to U^*$, so that only the last term, proportional to $Im(J_{ij}^{\beta\alpha})$, changes sign. We are finding the result analogous to the one that we have discussed in the case of the quarks. CP invariance, which would imply $P(\nu_\alpha \to \nu_\beta) = P(\bar\nu_\alpha \to \bar\nu_\beta)$, is violated to the extent that U is complex. Furthermore the complex values of U must be intrinsic, i.e. non eliminable by a re-phasing of the lepton fields: $Im(J_{ij}^{\beta\alpha})$ is indeed a rephasing-invariant measure of CP violation in the lepton sector, analogous to the Jarlskog invariant defined in equation (5.21). Incidentally, notice that a re-phasing of the neutrino fields does not change anything in equation (8.7), which means, following what was said in Section 7.2, that oscillation experiments cannot distinguish between Dirac and Majorana neutrinos.

It can be explicitly checked that the oscillation probability (8.7) enjoys the following properties:

- Conservation of probabilities:

$$\sum_\beta P(\nu_\alpha \to \nu_\beta) = \sum_\beta P(\bar{\nu}_\alpha \to \bar{\nu}_\beta) = 1 \qquad (8.8)$$

- CPT invariance:

$$P(\nu_\alpha \to \nu_\beta) = P(\bar{\nu}_\beta \to \bar{\nu}_\alpha) \qquad (8.9)$$

[*Problem 8.1.1: Prove explicitly these equations.*]

It is useful to specialize equation (8.7) to the two neutrino case, where U is, without loss of generality, a real orthogonal matrix parametrized by an angle θ. In this case the oscillation probability, say, from an electron to a muon neutrino is

$$\begin{aligned} P(\nu_e \to \nu_\mu) &= \sin^2 2\theta \sin^2 \frac{L \Delta m_{12}^2}{4E} \\ &\approx \sin^2 2\theta \sin^2 (1.27 \frac{L(\text{km}) \Delta m_{12}^2(\text{eV}^2)}{E(\text{GeV})}). \end{aligned} \qquad (8.10)$$

All other oscillation probabilities are obtained from this one by unitarity and CP or CPT invariance.

8.2. Neutrino propagation in matter

The coherent forward scattering of neutrinos propagating in matter leads to a modification of neutrino oscillations that is crucial to take into account if one wants to describe correctly the experimental results. The coherent sum of the scattered neutrino wave functions in the forward direction from every scattering center in matter changes the propagation of a relativistic neutrino by introducing, not surprisingly, an effective *refraction index*. This is totally analogous to what happens in optics, including the fact that the refraction index depends in general not only on the material that is crossed by the wave but also on the type of wave itself (e.g., in optics, the type of polarization).

Performing the sum over the scattered wave functions leads to the refraction index

$$n_\nu = 1 + \frac{2\pi}{p^2} \Sigma_i N_i f_i^\nu(0), \qquad (8.11)$$

where p is the neutrino momentum (of negligible mass), N_i is the density of the scattering center (i for electrons, nucleons or even the same neutrinos in a supernova) and $f_i^\nu(0)$ is the forward scattering amplitude of the neutrino ν over the same scattering center. [*Problem 8.2.1: Derive equation* (8.11).]

The scattering amplitude is either mediated by Z or W exchanges. We are only interested in effects that distinguish the different type of neutrinos. The forward scattering is coherent for neutrinos of definite lepton flavour, ν_α, $\alpha = e, \mu, \tau$. Since matter is composed of electrons (rather than μ or τ), ν_e interacts differently from $\nu_{\mu,\tau}$, giving rise to a flavour-dependent refraction index. In fact the scattering of any ν_α on electrons or quarks mediated by Z-exchange is the same for all flavours and therefore can be neglected. The only relevant effect comes therefore from νe scattering mediated by the W exchange. In a background composed by non-relativistic and non-polarized electrons (like on earth and, to a very good approximation, on the sun) the forward scattering amplitude for the electron neutrino is

$$f_e^\nu(0) = -\frac{G_F p}{\sqrt{2}\pi}, \tag{8.12}$$

and it changes sign for the electron anti-neutrino. [*Problem 8.2.2: Derive equation* (8.12).]

Working to first order in the neutrino masses and in $(n_\nu - 1)$, the evolution of the neutrino amplitude (8.2) in flavour space is therefore described, up to an overall phase e^{ipx}, by

$$-i\frac{d\mathcal{A}}{dx} = \mathcal{H}\mathcal{A}, \tag{8.13}$$

where

$$\begin{aligned}\mathcal{H}_{\beta\alpha} &= p(n_\nu - 1)\delta_{\beta e}\delta_{\alpha e} - \sum_i U_{\beta i}\frac{m_i^2}{2E}(U_{\alpha i})^* \\ &\approx -\sqrt{2}G_F N_e \delta_{\beta e}\delta_{\alpha e} - \sum_i U_{\beta i}\frac{m_i^2}{2E}(U_{\alpha i})^*.\end{aligned} \tag{8.14}$$

It is this equation that has to be solved to get the oscillation amplitudes in matter rather than in vacuum. The special novelty with respect to the propagation in vacuum is the x-dependence, through N_e, of the first term in the right hand side of (8.14). This is especially important if: i) some of the diagonal elements of $\mathcal{H}$ become degenerate at some point along the neutrino trajectory from the source to the detector; ii) the variation with x of $\mathcal{H}$ is slow enough that the neutrino state in flavour space can follow *adiabatically* the evolution of the eigenvalues of $\mathcal{H}$ itself. In such a case one gets what is called a *resonant* transition between different flavour states, where the matter effects are particularly relevant.

In this description of the neutrino propagation in matter we have never had to make the distinction between Majorana and Dirac neutrinos: they behave exactly in the same way. Even in presence of matter, oscillations do not allow to distinguish between these two types of neutrinos.

Finally one obtains the evolution equation for anti-neutrinos from (8.13), (8.14) by $U \to U^*$ and by changing the sign of the refraction-index term. As already said, the first is a genuine CP-violating effect, whereas the second is due to the breaking of CP by the background of ordinary matter. To prove the existence of CP violation in the lepton sector requires the experimental separation of these two effects.

8.3. Current determination of neutrino masses and mixings

At the time of writing these Lectures there are two well established *neutrino anomalies*, which can be clearly interpreted as due to neutrino oscillations: the *atmospheric* anomaly and the *solar* anomaly. In terms of three active neutrinos, either Dirac or Majorana, both anomalies can be described as effective two neutrino oscillations, each with its own squared mass difference and mixing angle, like in equation (8.10), with a third mixing angle being compatible with zero and actually mostly constrained by an independent reactor experiment, CHOOZ. With this in mind, the present information on neutrino masses and mixings from oscillations in a three light neutrino scheme are summarized in Table 8.1.

Table 8.1. *Summary of present information on neutrino masses and mixings from oscillation data.*

Oscillation parameter	Central value	99% CL range
solar mass splitting	$\Delta m^2_{12} = (8.0 \pm 0.3)10^{-5}$ eV^2	$(7.2 \div 8.9)10^{-5}$ eV^2
atmospheric mass splitting	$|\Delta m^2_{23}| = (2.5 \pm 0.2)10^{-3}$ eV^2	$(2.1 \div 3.1)10^{-3}$ eV^2
solar mixing angle	$\tan^2 \theta_{12} = 0.45 \pm 0.05$	$30° < \theta_{12} < 38°$
atmospheric mixing angle	$\sin^2 2\theta_{23} = 1.02 \pm 0.04$	$36° < \theta_{23} < 54°$
'CHOOZ' mixing angle	$\sin^2 2\theta_{13} = 0 \pm 0.05$	$\theta_{13} < 10°$

Since neutrino oscillations are at present the only positive evidence for neutrino masses, this information is alternatively pictured in Figs. 8.1 and 8.2 . Only knowing the two independent Δm^2 and not their sign, two possible neutrino spectra are compatible with Table 8.1: the *normal* or the *inverted* spectra, with an obvious meaning of the words. For the same reason, the center of mass of the neutrinos is equally undetermined. Oscillations of three active neutrinos only say at the moment that at least one neutrino is heavier than about $|\Delta m_{23}^2|^{1/2} \approx 5 \cdot 10^{-2} \text{eV}$. We have seen in the previous Lecture how β-decay experiments, in conjunction with the information in Table 8.1, allow to set a limit on the mass of every active neutrino of about $1 \div 2$ eV.

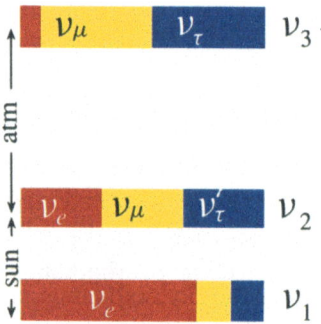

Figure 8.1. The so called *normal* spectrum, with the smallest splitting among the two lighter neutrinos. In different colours are also indicated the compositions of the three mass eigenstates in terms of ν_e, ν_μ, ν_τ.

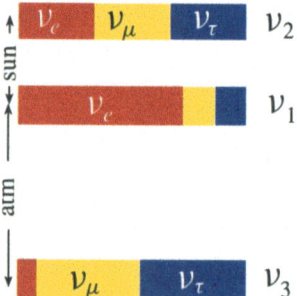

Figure 8.2. The so called *inverted* spectrum, with the smallest splitting among the two heavier neutrinos. In different colours are also indicated the compositions of the three mass eigenstates in terms of ν_e, ν_μ, ν_τ.

Chapter 9
The naturalness problem of the Fermi scale

9.1. The Standard Model as a prototype effective theory

The Standard Model, as described so far, is the *renormalizable* theory invariant under the gauge group $SU(3) \times SU(2) \times U(1)$ with three 16-plets of Weyl spinors transforming as in (1.16) and a single scalar doublet of hypercharge 1/2. As remarked in Appendix D for a general renormalizable theory, this invites to interpret the Standard Model as the low energy remnant of a more complete theory involving also higher physical scales than $G_F^{-1/2}$, denoted as in Lecture 4 by Λ_{NP}: the *physical* Standard Model as an *effective* field theory probably valid only up to some maximal energy scale $E_{\max}$ and closer to reality than the *renormalizable* Standard Model. $E_{\max}$ would have to be sufficiently smaller than Λ_{NP}, *i.e.* the smallest of the extra physical scales occurring in the extended theory.

This is good news and bad news at the same time. The good news is that the status of the Standard Model is enhanced: there are infinitely many theories whose experimental manifestations involving a maximal scale $E_{\max}$ are practically indistinguishable from it. Given the inherent limitations on $E_{\max}$, which change with time but are anyhow present at any given time, like now, this is conceptually very satisfactory. The bad news is that this makes it harder to guess which physical theory, if any, (partly) completes the Standard Model at higher energies. To a large extent, this is the main difficulty encountered in trying to go *beyond the Standard Model*.

This same logical framework, if accepted, gives however a clue towards possibly solving this difficulty, or at least allow to address it. A property of the more complete theory one is looking for, if any, is that it should reproduce in its low energy spectrum, well below Λ_{NP}, the particle content of the Standard Model. This requirement is trivially satisfied, by construction, for the gauge bosons, since the gauge group must contain $SU(3) \times SU(2) \times U(1)$. In fact, if the gauge group is larger, the generators of the extra symmetries must be broken at sufficiently high energy not to conflict with experiments. The same requirement is naturally fulfilled by the fermions as well, given that their masses also vanish in the $SU(3) X SU(2) X U(1)$ symmetric limit (except for the right handed neutrinos, which are irrelevant in this context). Furthermore, after breaking of the electroweak group, the fermion masses are controlled by parameters, the Yukawa couplings, that break the flavour symmetries of the minimal gauge Lagrangian. In all this, we are evidently assuming that the Fermi scale, or v, is sufficiently decoupled from Λ_{NP}, but this is precisely the point. Given equation (3.20), what is it, at last, that keeps the Higgs doublet light relative to Λ_{NP}? This is the possible clue to understand some key property of the theory that is supposed to extend the Standard Model. It goes under the name of *naturalness problem of the Fermi scale*.

9.2. Expanding in operators of higher dimension

Before trying to address this problem we ask if anything can be said on Λ_{NP} itself. Since the most precise tests of the Standard Model are successful, we can only hope of setting a lower bound on Λ_{NP}. To this end we use the fact that in the Standard Model as an effective theory the effects of new physics can be generally accounted for by introducing suitable higher dimensional operators weighted by appropriate inverse powers of Λ_{NP} (see Appendix D)

$$\mathcal{L}_{\text{SM}}^{\text{eff}} = \mathcal{L}_{\text{SM}} + \Sigma_{n,i} \frac{c_{n,i}}{\Lambda_{NP}^n} O_i^{(n+4)}. \tag{9.1}$$

Let us pause for a moment on this equation and, in particular, on the symmetries that the $O_i^{(n+4)}$ must respect. For sure the gauge symmetry of the Standard Model is one: a single non gauge invariant operator would destroy the consistency of the entire theory. Should they respect other symmetries? The physical theory that (9.1) is supposed to approximate may contain several scales, of which Λ_{NP} is the lowest. To set a credible lower bound on Λ_{NP} we have to use the most conservative assumption about the symmetries of the relevant $O_i^{(n+4)}$. We therefore pretend that they respect the full symmetry of the minimal gauge Lagrangian, $\mathcal{L}_{\text{min}}$, as discussed in Lecture 2. Since $\mathcal{L}_{\text{SM}}$ itself does not respect these symmetries, there will certainly be higher dimensional operators in (9.1) that do not respect them either, but their coefficients can be sufficiently suppressed, as we shall discuss in a while in an explicit example.

The list of operators of the lowest possible dimension that respect the symmetries of $\mathcal{L}_{\text{min}}$ is long. An interesting subset, complete if one does not include fermions, is

$$\mathcal{O}_{WB} = gg'(H^+\sigma^a H)W^a_{\mu\nu}B_{\mu\nu} = -2\frac{g'}{g}m_W^2 W^3_{\mu\nu}B_{\mu\nu} + \cdots, \quad (9.2)$$

$$\mathcal{O}_H = |H^\dagger D_\mu H|^2 = \left(\frac{m_W^2}{g^2}\right)^2 (gW^3_\mu - g'B_\mu)^2 + \cdots, \quad (9.3)$$

$$\frac{(D_\rho W^a_{\mu\nu})^2}{2} = \frac{(\partial_\rho W^a_{\mu\nu})^2}{2} + \cdots, \quad (9.4)$$

$$\frac{(D_\rho B_{\mu\nu})^2}{2} = \frac{(\partial_\rho B_{\mu\nu})^2}{2} + \cdots, \quad (9.5)$$

where the dots stand for terms not contributing to the vacuum polarization amplitudes. By comparison with the definitions given in equations (4.12, 4.15, 4.19), it is straightforward to see that each of these operators is in one to one correspondence with the form factors $\hat{S}, \hat{T}, W$ and Y respectively. Their presence in the effective Lagrangian (9.1) contributes to these form factors as

$$\begin{aligned}
\hat{S} &= 4\frac{m_W^2}{\Lambda_{NP}^2}c_{WB}, \\
\hat{T} &= -\frac{2}{g^2}\frac{m_W^2}{\Lambda_{NP}^2}c_{HH}, \\
W &= 2\frac{m_W^2}{\Lambda_{NP}^2}c_{WW}, \\
Y &= 2\frac{m_W^2}{\Lambda_{NP}^2}c_{BB},
\end{aligned} \quad (9.6)$$

where we have renamed the dimensionless coefficients $c_{n,i}$ in front of each operator using a self-explanatory notation. [*Problem 9.2.1: Prove these equations.*]

If we now compare these last equations with the experimental constraints summarized in Table 4.2, taking the dimensionless coefficients $c_{n,i}$ equal to unity gives a lower bound on Λ_{NP} at 95% C.L. of about 5 TeV from $\hat{S}$ or $\hat{T}$, whereas the limits on the two other form factors give a slighter weaker bound. Similar significant bounds are also obtained from the presence of some effective 4-fermion interactions, like, *e.g.*, the one involving 4 lepton doublets, summed over the flavour indices,

$$\mathcal{O}_{LL} = \frac{1}{2}(\bar{L}\gamma_\mu\sigma^a L)^2, \quad (9.7)$$

that correct the μ-decay amplitude. Note that, using the equations of motion derived from $\mathcal{L}_{SM}$, it is possible to recast (9.7) into a combination of operators only contributing to the vacuum polarization amplitudes of the gauge bosons plus new effective couplings of the gauge bosons to the quarks and the neutrinos, which are relatively less constrained experimentally. Specifically (9.7) is equivalent to a combinations of contributions to $\hat{S}, \hat{T}, \hat{U}, V$ and W with exactly correlated coefficients. [*Problem 9.2.2: Obtain explicitly these coefficients.*]

9.3. Minimal Flavour Violation

What if one had considered operators breaking the flavour symmetries of the minimal gauge Lagrangian? Had we used unsuppressed dimensionless coefficients in front of them, we would have been led to much stronger bounds on Λ_{NP}, but, as already mentioned, this would have been an unreasonable conclusion, since the flavour symmetries in $\mathcal{L}_{SM}$ are only broken by relatively small Yukawa couplings. Nevertheless, effective operators breaking the flavour symmetries have to be expected, which calls for an estimate of their possible impact on the considerations of this lecture. A useful concept to this end is the one of Minimal Flavour Violation.

Let us consider the full $SU(3)^5 X U(1)^4$ flavour symmetry of the Standard Model before switching on the Yukawa Lagrangian, as discussed in Lecture 2. We can attribute fictitious transformation properties under this symmetry group to the various Yukawa couplings, the numerical λ matrices appearing in (2.5), in such a way that a fictitious $SU(3)^5 X U(1)^4$ symmetry is kept intact even in their presence, like we do, *e.g.*, by spurion techniques to check the consistency of a calculation. We assume that all the flavour breaking operators present in $\mathcal{L}_{SM}^{eff}$ are weighted by dimensionless coefficients of order unity apart from the minimum number of λ matrices that are needed for them to respect the fictitious $SU(3)^5 X U(1)^4$ symmetry so defined. It should be clear why this assumption is called Minimal Flavour Violation.

Two relevant examples of this are

$$\mathcal{L}_{eff}^{MFV} = \frac{c_{\Delta F=2}}{\Lambda_{NP}^2} \frac{1}{2} (\bar{Q}(\lambda^U)^*(\lambda^U)^T \gamma_\mu Q)^2 \\ + \frac{c_{b \to s\gamma}}{\Lambda_{NP}^2} H^+ (Q^T \lambda^U (\lambda^U)^+ \lambda^D \sigma_{\mu\nu} d) B^{\mu\nu} + \text{h.c.} \quad (9.8)$$

in the notation of Lecture 2. If, in accordance with the stated assumption, we take $|c_{\Delta F=2}| = |c_{b \to s\gamma}| = 1$, the first operator, either from B or from K mixing, and the second operator from $BR(b \to s\gamma)$ lead to lower bounds on Λ_{NP} of about 5 TeV, completely analogous to the ones obtained in the previous Section from flavour-conserving physics. [*Problem 9.3.1: Show the invariance of the operators in (9.8) under the fake flavour symmetry.*]

9.4. The naturalness scale of the Standard Model

Let us go back now to the naturalness problem of the Fermi scale, which is why the Higgs boson is light. At least to attack it, it is necessary that the Higgs boson mass be calculable in terms of some other physical scale of the theory that properly extends the Standard Model. While this is obvious, it is also clearly impossible to get an expression for the Higgs boson mass without specifying which extension of the Standard Model we are talking about. Nevertheless, and not surprisingly, in all known examples a contribution to this mass comes from the radiative loops of the Standard Model particles suitably cut-off at some of the new scales that characterize the extended theory.

To estimate this contribution, we consider the one-loop corrections to the effective potential of the Higgs field

$$\Delta V(h) = \frac{1}{2} \int \frac{d^4 p}{(2\pi)^4} STr(\log(p^2 + M^2(h))) \qquad (9.9)$$

where the supertrace is defined by $STr \equiv Tr(-1)^{2S}$ and S is the spin of the particle of h-dependent mass $M(h)$ circulating in the loop. The integral has also been rotated to euclidian 4-momentum in the standard way. Cutting the integral at $p^2 < \Lambda^2$ and expanding for large Λ, the leading h-dependent term of the effective potential is therefore a contribution to the parameter μ^2 in (2.6)

$$-\delta\mu^2 \frac{h^2}{2} = \frac{\Lambda^2}{32\pi^2} STr M^2(h), \qquad (9.10)$$

which can be expressed in terms of the masses of the various particles. From Section 3.3 one has

$$\begin{aligned} M_V &= m_V \frac{h}{\sqrt{2}v}, \\ M_f &= m_f \frac{h}{\sqrt{2}v}, \\ M_h &= \sqrt{\frac{3}{2}} m_h \frac{h}{\sqrt{2}v}, \\ M_{\pi_i} &= \sqrt{\frac{1}{2}} m_h \frac{h}{\sqrt{2}v}, \end{aligned} \qquad (9.11)$$

so that in view of (3.20), keeping only the top contributions from the fermions,

$$\delta m_h^2 = 2\delta\mu^2 = \frac{3\Lambda^2}{16\pi^2 v^2}(4m_t^2 - 2m_W^2 - m_Z^2 - m_h^2). \tag{9.12}$$

As said, in all explicit examples which address the naturalness problem of the Higgs boson mass, this contribution is indeed present with the different terms generally cut-off by different uncorrelated physical mass scales instead of a single Λ as in (9.12). Without unwarranted cancellations, it is therefore reasonable to define a naturalness scale of the Standard Model, Λ_{nat}, as the scale at which it is necessary to cut-off the dominant of these terms to avoid that it exceeds the physical Higgs boson mass by some moderate amount $\sqrt{\Delta}$. For a Higgs boson mass well below about 350 GeV, as apparently implied by the precision tests discussed in Lecture 4, the top contribution dominates and leads to

$$\Lambda_{\text{nat}} < \frac{2\pi}{\sqrt{3}} \frac{v}{m_t} m_h \sqrt{\Delta} \approx 400 \text{ GeV} \frac{m_h}{115 \text{ GeV}} \sqrt{\Delta}. \tag{9.13}$$

$1/\Delta$ is the percentage of cancellation that could take place between the top contribution and any other possible term in the complete expression for the mass squared of the Higgs boson. For an accidental cancellation, the larger is Δ the weaker is the claim that the lightness of the Higgs boson is understood.

A possible confusion about this way of reasoning must be avoided. Λ_{nat} has nothing to do with the regularization cut-off of the Standard Model or of whatever theory one is talking about. This cut-off has been sent to infinity, or at least well above any physical scale, as one should in field theory. As repeatedly said, equation (9.10) only serves as a way to estimate the physical contributions present in the extended theory to the Higgs boson mass.

9.5. The *little hierarchy* problem

If Λ_{nat} is identified with some new physical scale in the extended Standard Model, the upper bound in (9.13) is in conflict with the lower bound on Λ_{NP} derived in the previous Sections, unless one invokes an accidental cancellation at about 1% level. Such is indeed the square of the ratio of Λ_{nat}, equation (9.13), to the minimum Λ_{NP}, of about 5 TeV, for $\Delta = 1$ and a Higgs boson mass in the 100 GeV range. Since this is not quite satisfactory, we call it the *little hierarchy* problem, to distinguish it from the problem of understanding the smallness of the Higgs mass relative to some very high scale identified a priori, like the Planck scale or the Grand Unification scale, and called *hierarchy* problem tout court.

The little hierarchy problem has direct phenomenological relevance. If the Fermi scale is natural in the sense defined above, should one not have already seen the physics that makes it natural at all? There are several caveats that do not make it possible to answer this question at the time of writing these lectures. At least two are worth mentioning. The lower bound on Λ_{NP} of several TeVs rests on taking the dimensionless coefficients $c_{n,i}$ in (9.1) equal to unity, which is inappropriate if the new physics is perturbative and contributes at loop level only. Furthermore it is possible that the physical interpretation of the precision tests in terms of a light Higgs boson, although consistent with the data and necessary in the Standard Model, (see equation (4.17)), is not the correct one. Such an interpretation, on the other hand, lies behind the little hierarchy problem, since Λ_{nat} in equation (9.13) is proportional to m_h: a Higgs boson in the 400 ÷ 500 GeV range, rather than close to 100 GeV, would relax the bound on Λ_{nat} in a non negligible way. The experiments at the Large Hadron Collider are supposed to shed light on these issues.

Chapter 10
The main drawback of the Standard Model

10.1. Gauge anomalies and charge quantization

As recalled in Appendix A, in any gauge theory like the Standard Model with a $U(1)$ among the gauge-group factors, the charges of the various matter representations under such a factorized $U(1)$ are unconstrained by gauge invariance itself. This is at variance with the eigenvalues of a diagonal generator belonging to a non-abelian gauge-group factor or with the gauge couplings to the non-abelian gauge bosons felt by different matter representations of a non-abelian gauge group: in a simple non-abelian gauge group all of these quantities are fixed in terms of a single coupling constant which determines also the strength of the self-interactions among the gauge bosons. This remarkable property of non-abelian gauge groups has of course been extensively used in the previous Lectures to compare the Standard Model with experiment. However, as already remarked in Lecture 1, it has been apparently necessary to fix by hand the hypercharges of each irreducible representation of the Standard Model gauge group to match, after electroweak symmetry breaking, the electric charges of the various matter fields. Even insisting on assigning these charges in such a way as to keep parity conserved by the strong and the electromagnetic interactions, we could have taken an irrational number for the relative charge of quark and leptons and still be consistent with the classical $SU(3) \times SU(2) \times U(1)$ gauge invariance. Given the observed equality of the proton and the electron charges, within more that 20 digits, this is what I call the main drawback of the Standard Model.

The attentive reader may have noticed that I talk of consistency with the *classical* gauge symmetry. In fact, as recalled in Appendix G, a classical gauge symmetry may turn out to be *anomalous*, i.e. broken by perturbative quantum effects. For a consistent theory, however, the gauge symmetry must be non anomalous, which requires that the totally symmetric triple product of any of its generators has vanishing trace

$$Tr(T^a T^b T^c)_S = 0, \qquad (10.1)$$

over the full set of fermion fields, all taken left handed as usual. Is the Standard Model free of anomalies? Furthermore, can it be that the requirement of absence of gauge anomalies fixes the proton to electron charge in the desired way?

To answer these questions, it is useful to examine a preliminary statement. Consider any $SU(3) \times SU(2) \times U(1)$ gauge theory with a fermion representation made of doublets and/or singles under SU(2) and real under colour and electromagnetism (the subgroup $SU(3) \times U(1)_Q$ where, as usual, $Q = T_3 + Y$). Such a theory is free of gauge anomalies if and only if

$$Tr[(T_3)^2 Q] = 0, \tag{10.2}$$

where T_3 is a generator of $SU(2)$. In words the condition is that the charges of the $SU(2)$-doublets have to add up to zero. This statement is easy to prove by making use of Appendix B. In particular a real representation of a gauge group has no anomaly. Furthermore $SU(2)$ has only real representations. [*Problem 10.1.1: Prove these statements.*]

The absence of gauge anomalies in the Standard Model is now readily checked: it results from

$$Tr[(T_3)^2 Q] = \frac{1}{4} \times 3 \times \left(\frac{2}{3} - \frac{1}{3}\right) + \frac{1}{4} \times (0 - 1) = 0, \tag{10.3}$$

where the first term comes from the quarks and the second from the leptons (for each generation). The absence of anomalies boils down to the cancellation of charges between quarks and leptons. Furthermore we also seem to have remedied to the problem left to us from classical gauge invariance: charge non quantization. Let us examine this statement more in detail.

We can ask the following question. What would be the simplest fully chiral representation of the $SU(3) \times SU(2) \times U(1)$ gauge group for the matter fermions, that would give an anomaly free theory? By *simplest* we mean with the least number of Weyl spinors. The fifteen-plet of Weyl spinors that make one generation of the Standard Model, excluding the right handed neutrino, fails by little to be such a simplest representation. The simplest representation is in fact a charge 1/2 quark, i.e. in the notation of Lecture 1,

$$\Psi_{1/2} = (\widehat{Q}(3, 2)_0, \widehat{u}^c(\bar{3}, 1)_{-1/2}, \widehat{d}^c(\bar{3}, 1)_{1/2}), \tag{10.4}$$

which does not seem to exists in nature. [*Question 10.1.1: In what precise sense a charge 1/2 lepton is unsuitable here?*]

The fifteen-plet of the Standard Model is the next to simplest anomaly-free chiral representation of $SU(3) \times SU(2) \times U(1)$ and, for this, it is crucial that charge is quantized, as shown in (10.3). However it suffices to add a single Weyl spinor to get the cancellation of all the anomalies, while allowing for an arbitrary charge, as follows:

$$\Psi_y = \left(\tilde{Q}(3,2)_y, \tilde{L}(1,2)_{-3y}, \tilde{u}^c(\bar{3},1)_{-y-1/2}, \tilde{d}^c(\bar{3},1)_{-y+1/2}, \right. \\ \left. \tilde{e}^c(1,1)_{3y+1/2}, \tilde{N}(1,1)_{3y-1/2}\right) \quad (10.5)$$

This is the same number of Weyl spinors as one generation of Standard Model fermions, which was given in Lecture 1, with the inclusion of the right-handed neutrino. In fact one reproduces the basic representation of the Standard Model (1.16) by setting $y = 1/6$. In this sense the relation of charge quantization to anomaly cancellation gets weakened.

10.2. The unification way

Unlike the charges under a factorized $U(1)$, a charge included in the generators of a non abelian group is quantized by the commutation rules of the algebra, like, e.g., the eigenvalues of T_3 in $SU(2)$. To embed $SU(3) \times SU(2) \times U(1)$ into a simple group looks therefore as a neat way to solve the charge quantization problem. Let us see how this can be done within a $SU(N)$ group.

The number of generators that commute among each other in $SU(3) \times SU(2) \times U(1)$ (or the *rank* of the group) is 4: 2 in $SU(3)$, 1 in $SU(2)$ and 1 in $U(1)$. Since $SU(N)$ has rank $N - 1$, $SU(5)$ is the smallest candidate to do the job. It turns out that it is also the single group with minimal rank. Furthermore there is an obvious single way of embedding $SU(3) \times SU(2) \times U(1)$ in $SU(5)$. It amounts to see the $SU(5)$-generators restricted to the $SU(3) \times SU(2)$ subgroup as

$$\begin{pmatrix} \frac{1}{2}\lambda^i_{3\times 3} & \\ \hline & \frac{1}{2}\sigma^a_{2\times 2} \end{pmatrix} \quad (10.6)$$

where $\lambda^i_{3\times 3}/2, i = 1, \ldots, 8$ and $\sigma^a_{2\times 2}/2, a = 1, 2, 3$ are the generators of $SU(3)$ and $SU(2)$ respectively. Consequently

$$Y \propto \begin{pmatrix} 1_{3\times 3} & \\ \hline & -\frac{3}{2}1_{2\times 2} \end{pmatrix} \quad (10.7)$$

has to be the necessarily traceless hypercharge $U(1)$-generator, up to an overall normalization factor. As anticipated, we have automatically achieved charge (or rather hypercharge) quantization, even though we have still to see if the eigenvalues, so obtained, are suitable for physics.

Under which $SU(5)$-representation can the fermions transform? To see this we have to decompose the $SU(5)$ representations under $SU(3) \times SU(2) \times U(1)$. From (10.6) the fundamental of $SU(5)$ decomposes as

$$\underline{5} = (\mathbf{3}, \mathbf{1})_y \oplus (\mathbf{1}, \mathbf{2})_{-3/2y} \tag{10.8}$$

in the notation of Lecture 1, with the hypercharge still not normalized. Therefore the conjugate representation $\bar{5}$ is suited for containing a d^c and an L, with $y = -1/3$. We still have to find room for at least 10 more Weyl spinors: Q, u^c and e^c, with an obvious candidate in the 10-dimentional representation of $SU(5)$, i.e. the representation of next dimensionality.

The $\underline{10}$ of $SU(5)$ transforms as the antisymmetric product of two $\underline{5}$'s. Therefore, from (10.8) and $y = -1/3$,

$$\begin{aligned}\underline{10} = (\underline{5} \times \underline{5})_A &= ((\mathbf{3} \times \mathbf{3})_A, \mathbf{1})_{-2/3} \oplus (\mathbf{1}, (\mathbf{2} \times \mathbf{2})_A)_1 \oplus (\mathbf{3}, \mathbf{2})_{1/6} \\ &= (\bar{\mathbf{3}}, \mathbf{1})_{-2/3} \oplus (\mathbf{1}, \mathbf{1})_1 \oplus (\mathbf{3}, \mathbf{2})_{1/6}\end{aligned} \tag{10.9}$$

which contains, miraculously enough, precisely the leftover multiplets, except for the right handed neutrino which will necessarily have to be an overall $SU(5)$ singlet. It is important to realize that not only the $SU(3) \times SU(2)$ properties but also the hypercharges of the various multiplets did not have to come out right a priori. The correct hypercharge quantization is therefore achieved by embedding the 16-plet of Weyl spinors of one generation of matter into a $(\underline{10} \oplus \bar{\underline{5}} \oplus \underline{1})$ of $SU(5)$.

Appendix A
General structure of a gauge theory

In an arbitrary gauge field theory, the form of the Lagrangian $\mathcal{L}_g$ is determined by:

1. The gauge group $\mathcal{G}$, with generators T_A, $A = 1, 2, ..., N$, satisfying a characteristic algebra

$$[T_A, T_B] = i f_{ABC} T_C; \qquad (A.1)$$

2. The representation r_Ψ under $\mathcal{G}$ of the fermionic fields Ψ_α, $\alpha = 1, 2, \ldots, N_\Psi$, all conventionally taken as left-handed Weyl spinors;
3. The representation r_ϕ under $\mathcal{G}$ of the scalar fields ϕ_a, $a = 1, 2, ..., N_\phi$.

If we restrict ourselves to local monomials in $\mathcal{L}_g$ of mass dimension less then or equal to four, for which there is a good reason (see Appendix D) Poincare' invariance and gauge invariance require $\mathcal{L}_g$ to be of the form, up to possible field redefinitions [*Question A.1: Why this qualification?*] and neglecting for the time being total derivatives,

$$\mathcal{L}_g = \mathcal{L}_{\min} - \left(\frac{1}{2}\Psi^T M \Psi + \phi \Psi^T \lambda_1 \Psi + \phi^+ \Psi^T \lambda_2 \Psi + \text{h.c.}\right) - V(\phi) \quad (A.2)$$

where $V(\phi)$ is a gauge invariant hermitian polinomial of degree up to four in the fields ϕ_a and

$$\mathcal{L}_{\min} = -1/4 F^A_{\mu\nu} F^A_{\mu\nu} + i\bar{\Psi}\slashed{D}\Psi + |D_\mu \phi|^2 \qquad (A.3)$$

$$F^A_{\mu\nu} = \partial_\mu A^A_\nu - \partial_\nu A^A_\mu + g f_{ABC} A^B_\mu A^C_\nu \qquad (A.4)$$

$$D_\mu = \partial_\mu - ig A^A_\mu T^A \qquad (A.5)$$

Since the notation is concise, although also precise, several comments are in order:

1. For any index A there are corresponding vector bosons A_μ^A, with 3-linear and 4-linear self-interactions, and interacting with the fermion and the scalar fields through the currents

$$J_\mu^A(\Psi) = \bar\Psi \gamma_\mu T^A \Psi, \quad J_\mu^A(\phi) = \phi^+ T^A \partial_\mu \phi - \partial_\mu \phi^+ T^A \phi \quad (A.6)$$

   All these gauge interactions are contained in the minimal gauge Lagrangian $\mathcal{L}_{\min}$.
2. Except for the index A in $F_{\mu\nu}^A$, all the other gauge indices are left implicit. Hence M and $\lambda_{1,2}$ are in general (sometimes necessarily vanishing) matrices in gauge-group space, so that invariance under $\mathcal{G}$ is guaranteed. Similarly D_μ is a matrix in gauge-group space, generally different when it acts on Ψ_α or ϕ_a.
3. Notice the appearance of a single coupling g in $\mathcal{L}_{\min}$. This is indeed forced by gauge invariance if the group $\mathcal{G}$ is simple and does not contain $U(1)$ factors. In general, for any simple non-abelian factor of $\mathcal{G}$ there is a single arbitrary constant, whereas the charges of the fermion and scalar fields under any $U(1)$-factor of the gauge group are all arbitrary (see Appendix G, though)
4. Also the Lorenz indices of the fermion fields are left understood. There is no ambiguity here, since a unique contraction of the Lorenz indices is possible in a fermion bilinear to get either a scalar or a vector. Notice also the notation for the fermion mass terms (or for the Yukawa couplings) in equation (A.2) which is different from the one generally adopted in QED or in QCD, where parity invariance makes it highly preferable to work with Dirac spinors. [*Problem A.1: Write the QED or the QCD Lagrangian in this notation.*]

Appendix B
Real and chiral representations of the gauge group $\mathcal{G}$

Under the group $\mathcal{G}$ the fermion multiplet Ψ transforms as

$$\Psi \Rightarrow e^{i\omega_A t_A}\Psi \qquad (B.1)$$

where t_A are the hermitian matrices that represent the generators T_A as acting on Ψ and ω_A are the x-dependent real parameters of the transformation.

A representation r_Ψ is said to be *real*, or *self-conjugate*, if there exists a matrix $M_{\alpha\beta}$ such that $M_{\alpha\beta}\Psi_\alpha\Psi_\beta$ is invariant under the above transformation. If no such matrix exists the representation is said to be *chiral*. The representation in question may be reducible or irreducible. If it is reducible and the matrix $M_{\alpha\beta}$ is non vanishing only when acting on some component of r_Ψ, it is this component which is called real, whereas the remaining part is chiral.

Given the representation r according to which Ψ transforms as in (B.1), χ transforms as the conjugate representation $\bar{r}$ if

$$\chi \Rightarrow e^{-i\omega_A t_A^*}\chi \qquad (B.2)$$

Since the t_A are hermitian, $\chi_\alpha\Psi_\alpha$ is invariant. This allows us to give an equivalent definition of a real representation (which explains the name). A representation is said to be real if there exists a unitary matrix U, such that

$$t_A^* = -Ut_A U^+ \qquad (B.3)$$

[*Problem B.1: Prove the equivalence of the two different definitions. Question B.1: Why a real representation is sometimes also called vector representation?*]

Appendix C
Spontaneous breaking of a gauge or a global symmetry

It is best to describe the general features of spontaneous symmetry breaking of a global or a local symmetry by considering their implementation in a scalar field theory

$$\mathcal{L}(\phi_a) = \frac{1}{2}(D_\mu \phi_a)^2 - V(\phi_a) \tag{C.1}$$

where ϕ_a is a multiplet of scalar fields, taken real for convenience, and $V(\phi_a)$ is an hermitian potential, invariant under a group G, which acts on the ϕ_a as

$$\phi \Rightarrow e^{-w^A t^A}\phi. \tag{C.2}$$

Since the ϕ_a are real, the t_A are real antisymmetric matrices. If the symmetry is local, the covariant derivative appears in (C.1)

$$D_\mu = \partial_\mu + gA_\mu^A t^A \tag{C.3}$$

together with the kinetic term for the gauge bosons.

The potential $V(\phi)$ is assumed to have a minimum for a non zero value of the $\phi_a = v_a$ (or $\phi = v$), where v is a constant vector. This configuration of the ϕ-fields minimizes the Hamiltonian of the system and is therefore called the *vacuum*. The generators t^a, $a = 1, ..., M$ which annihilate the vacuum, $t^a v = 0$, or leave it invariant, are called *unbroken* generators. Since they form a closed algebra, one says that a subgroup H of G is left unbroken. The remaining generators t^i, $i = M+1, ..., N$ are called *broken* generators.

The invariance of V under G implies that $\exp(-t^i \pi^i)v$ is also a mimimum of the potential degenerate with $V(v)$ for any π^i, which we could have had equally well picked up as reference vacuum without any change of the physics. Hence the $\pi^i(x)$, one for any broken generator t^i, are massless fields: they are called *Goldstone bosons*. [*Problem* C.1.1: *Show this formally by expanding the potential around the minimum* $\phi = v$.]

If the symmetry G is gauged, however, a symmetry transformation

$$\phi \Rightarrow e^{t^i \pi^i} \phi \tag{C.4}$$

(and correspondingly on the vectors) removes the π_i not only from V but also from the kinetic term in (C.1), so that the Goldstone boson fields disappear at all from the spectrum of the theory. This is called *going to the unitary gauge*. What happens of the corresponding degrees of freedom? By expanding the covariant kinetic term in (C.1) at the minimum configuration $\phi = v$ one finds a mass term for the gauge bosons

$$\Delta \mathcal{L}_m = \frac{1}{2} m^2_{AB} A^A_\mu A^B_\mu, \quad m^2_{AB} = g^2 (t^A v)_a (t^B v)_a. \tag{C.5}$$

Notice that the squared mass matrix m^2_{AB} is non vanishing only when restricted to the broken generators $A = i$ and that, in this sector, it is positive definite, since all its diagonal elements are positive. This shows that any vector A^i_μ corresponding to a broken generator acquires a mass. [*Problem C.1.2: Show the same result without eliminating the fields π by the gauge transformation* (C.4).] The appearance of the masses for the vector bosons while the Goldstone bosons are eliminated from the spectrum is called *Higgs phenomenon*. In practical calculations of Feynman diagrams it is often useful to use a gauge where the Goldstone bosons are propagating, even with a gauge-dependent non vanishing mass, although the corresponding poles always disappear at the end of the calculations from the physical spectrum.

Notice that in the ϕ_a there are always at least as many components (real!) as there are broken generators. This means that, once the *eaten up* Goldstone bosons are taken away, there are in general residual bosons in the physical spectrum, (as many as the components of ϕ minus the number of broken generators), whose masses depend on the details of V. They form as any other set of particles in the theory a representation of the unbroken group H.

Appendix D
Renormalizable theories and effective theories

Although the concept is in principle more general, here we call *renormalizable* a local quantum field theory involving scalars, spin-1/2 fermions and possibly gauge vectors (see Appendix A) if it contains all the monomials of degree 4 at most in the various fields, which are consistent with a given symmetry, global and/or local. The gauge symmetry must be non anomalous (see Appendix G). Such a theory can be perturbatively compared with experiments at all energies where its dimensionless couplings, which acquire an energy dependence from loop corrections, remain sufficiently small.

Renormalizable theories enjoy a remarkable property. Consider a theory respecting a given symmetry, part or all of it possibly gauged, which has in its low energy spectrum only scalars, spin-1/2 fermions and, if needed, the gauge vectors. For processes involving sufficiently low energies, any such theory is practically undistinguishable from the unique corresponding renormalizable theory, *i.e.* the renormalizable theory with the same low energy spectrum and the same symmetry.

Let us try to be more precise about what one means by *low energy spectrum* or *at sufficiently low energies*. Call $\mathcal{L}_{UV}$ the Lagrangian of the theory that we want to compare with the corresponding renormalizable one $\mathcal{L}_{ren}$. An interesting case is when $\mathcal{L}_{UV}$ has the same spectrum as $\mathcal{L}_{ren}$, but also contains terms of the form

$$\Delta \mathcal{L} = \Sigma_{n,i} \frac{c_{n,i}}{\Lambda^n} O_i^{(n+4)}, \tag{D.1}$$

where $O_i^{(n+4)}$ are operators of dimension $n+4$ with $n > 0$, the $c_{n,i}$ are dimensionless coefficients of order unity or less and Λ is a mass parameter. If the spectrum is also made of particles all lighter than Λ, then $\mathcal{L}_{UV}$ is undistinguishable from $\mathcal{L}_{ren}$ at energies $E \ll \Lambda$. For this to be the case, the spectra of the two theories do not have to coincide. It is necessary that the spectrum of $\mathcal{L}_{ren}$ corresponds only to the physical *infrared spectrum* of $\mathcal{L}_{UV}$, *i.e.* its spectrum at small energies relative to any other physical scale $\geq \Lambda$ occurring in $\mathcal{L}_{UV}$ itself.

We call *effective* a local field theory whose consequences are valid only up to some maximal energy scale $E^{\max}$. Effective field theories are generally non renormalizable and their Lagrangian contains operators of dimension higher than 4. From what was said in the previous paragraphs, it may be practically impossible to distinguish a renormalizable theory from an effective theory. A currently prevailing view is that all the theories we use to describe particle physics are in fact effective theories.

Appendix E
CP invariance

We can define a CP transformation of a fermion field ψ of given chirality as

$$\psi(\mathbf{x}, t) \to i\gamma_2\gamma_0\psi(-\mathbf{x}, t)^*. \tag{E.1}$$

Using the standard decomposition in creation and annihilation operators of the fermion field, it can be seen that this transformation amounts to

$$\text{particle}(\mathbf{p}, \mathbf{s}) \to \text{antiparticle}(-\mathbf{p}, -\mathbf{s}), \tag{E.2}$$

where $\mathbf{p}$ and $\mathbf{s}$ are the momentum and the spin.

From the definition (E.1) and the properties of the γ matrices it follows that under CP

$$\bar{\psi}\chi \to \bar{\chi}\psi \tag{E.3}$$

$$\bar{\psi}\gamma_\mu\chi \to -\bar{\chi}\gamma_{\tilde{\mu}}\psi \tag{E.4}$$

$$\bar{\psi}\sigma_{\mu\nu}\chi \to -\bar{\chi}\sigma_{\tilde{\mu}\tilde{\nu}}\psi, \tag{E.5}$$

where, in the right-hand-side of any of these equations, the spacetime argument is $(-\mathbf{x}, t)$ and every Lorentz index with a tilde requires an overall minus sign if it is a space index or a $+$ sign if it is a time index. Note that in eqs. (E.3, E.5) the chirality of ψ and χ is opposite, whereas in (E.4) it is the same. [*Problem E.1: Show equations* (E.3), (E.4), (E.5).]

From (E.4) it immediately follows that the free kinetic action $S = \int d^4x \mathcal{L}_{\text{kin}}^{\text{free}}$ of a fermion is invariant under CP since $\mathcal{L}_{\text{kin}}^{\text{free}} \to \mathcal{L}_{\text{kin}}^{\text{free}}(-\mathbf{x}, t)$. On the other hand, always from (E.4) a general gauge interaction transforms as (now Ψ is a multiplet of Weyl spinors)

$$A_\mu^a(\bar{\Psi}\gamma_\mu t^a \Psi) \to -(A_\mu^a)^{\text{CP}}(\bar{\Psi}\gamma_{\tilde{\mu}}(t^a)^T \Psi)(-\mathbf{x}, t) \tag{E.6}$$

which is invariant if

$$(A_\mu^a)^{\text{CP}} t^a = -A_{\tilde{\mu}}^a (t^a)^T(-\mathbf{x}, t). \tag{E.7}$$

This is the transformation law of the vectors under CP, which leaves also invariant the gauge kinetic term $F^a_{\mu\nu} F^a_{\mu\nu}$ after space-time integration since

$$F_{\mu\nu} = [D_\mu, D_\nu] \to -F^T_{\tilde\mu\tilde\nu}(-\mathbf{x}, t). \tag{E.8}$$

Notice on the contrary that $\epsilon_{\mu\nu\rho\sigma} F^a_{\mu\nu} F^a_{\rho\sigma}$ is odd under CP.

Appendix F
Weyl, Dirac and Majorana neutrinos

The general free Lagrangian for a set of Weyl spinors, χ_i, all conventionally taken left-handed, has the form, only restricted by Lorentz invariance,

$$\mathcal{L}_{\text{free}} = i\bar{\chi} A \bar{\partial} \chi - \left(\frac{1}{2}\chi^T M \chi + \text{h.c.}\right) \tag{F.1}$$

where A and M are matrices, respectively hermitian and symmetric. By field redefinitions A can be set equal to unity and M can be diagonalized, with real non negative eigenvalues.

With a single Weyl spinor, only two cases are possible:

- The free Lagrangian of a *Weyl neutrino*

$$\mathcal{L}_{\text{free}}^{\text{Weyl}} = i\bar{\chi} \bar{\partial} \chi \tag{F.2}$$

- The free Lagrangian of a *Majorana neutrino*

$$\mathcal{L}_{\text{free}}^{\text{Maj}} = i\bar{\chi} \bar{\partial} \chi - \left(\frac{1}{2}\chi^T M \chi + \text{h.c.}\right), \tag{F.3}$$

(where M is a single real positive mass). They are distinguished by the fact that only in the first case a charge counting the number of neutrinos may be conserved, depending on their interactions. A Weyl neutrino has definite helicity, opposite to the one of the charge conjugate state, χ^C. Due to the mass term, a Majorana neutrino does not have a fixed helicity and is often described by means of a self-conjugate Majorana spinor

$$\chi^M \equiv \chi + \chi^C. \tag{F.4}$$

In terms of χ^M, the free Lagrangian is

$$\mathcal{L}_{\text{free}}^{\text{Maj}} = \frac{1}{2} i \bar{\chi}^M \bar{\partial} \chi^M - \frac{M}{2} \bar{\chi}^M \chi^M. \tag{F.5}$$

Note that $(\bar{\chi}^M)^T$ and χ^M are proportional to each other up to a matrix acting on the spinor indices.

A particularly relevant case with two different Weyl spinors is when the 2×2 mass matrix admits a conserved charge, *i.e.* in a suitable basis M has only off-diagonal elements, $\chi^T M \chi = \chi_1 M_{12} \chi_2$, or, equivalently, it has degenerate eigenvalues, up to a phase. [*Problem* F.1: *Show this equivalence.*] In terms of a Dirac spinor

$$\chi^D \equiv \chi_1 + \chi_2^C, \tag{F.6}$$

the free Lagrangian for a *Dirac neutrino* is

$$\mathcal{L}_{\text{free}}^{\text{Dirac}} = i\bar{\chi}^D \slashed{\partial} \chi^D - M \bar{\chi}^D \chi^D. \tag{F.7}$$

Appendix G
Anomalies

We consider a theory described by a Lagrangian with gauge and global symmetries acting in particular on a multiplet of fermionic fields Ψ, all conventionally taken as left-handed Weyl spinors. In association with these symmetries there are currents which, restricted to the fermions, take the form

$$J_\mu^a(\Psi) = \bar{\Psi}\gamma_\mu T^a \Psi. \tag{G.1}$$

The index a can be associated either with a gauge or with a global current.

An important property of symmetries in field theory is that they may be broken by perturbative quantum effects. This arises when the radiative corrections to some Green functions make them inconsistent with the Ward Identities derived by standard procedures from the symmetries of the classical Lagrangian. A symmetry is called anomalous when such inconsistency is not removable by a suitable appropriate redefinition of the Lagrangian itself.

By inspection of the perturbative quantum corrections, it is possible to characterize if a symmetry is not anomalous by a neat criterium: it is sufficient and necessary that the totally symmetric triple product of its generators have vanishing trace,

$$Tr(T^a T^b T^c)_S = 0, \tag{G.2}$$

over the fermion multiplet Ψ for any a, b, c. One can have *global*, **mixed** or *gauge* anomalies, depending on whether the generators with non vanishing triple trace are all global, mixed or all gauge generators. Global and/or mixed anomalies are innocuous for the consistency of the theory or even welcome for their physical implications. Examples of both exist in the case of the Standard Model Lagrangian. On the contrary, gauge anomalies lead to an inconsistent theory and are therefore to be avoided.

LECTURE NOTES

This series publishes polished notes dealing with topics of current research and originating from lectures and seminars held at the Scuola Normale Superiore in Pisa.

Published volumes

1. M. TOSI, P. VIGNOLO, *Statistical Mechanics and the Physics of Fluids*, 2005 (second edition). ISBN 978-88-7642-144-0
2. M. GIAQUINTA, L. MARTINAZZI, *An Introduction to the Regularity Theory for Elliptic Systems, Harmonic Maps and Minimal Graphs*, 2005. ISBN 978-88-7642-168-8
3. G. DELLA SALA, A. SARACCO, A. SIMIONIUC, G. TOMASSINI, *Lectures on Complex Analysis and Analytic Geometry*, 2006.
ISBN 978-88-7642-199-8
4. M. POLINI, M. TOSI, *Many-body Physics in Condensed Matter Systems*, 2006. ISBN 978-88-7642-192-0
P. AZZURRI, *Problemi di Meccanica*, 2007.
5. R. BARBIERI, *Ten Lectures on the ElectroWeak Interactions*, 2007.
ISBN 978-88-7642-311-6

Volumes published earlier

G. DA PRATO, *Introduction to Differential Stochastic Equations*, 1995 (second edition 1998). ISBN 978-88-7642-59-1
L. AMBROSIO, *Corso introduttivo alla Teoria Geometrica della Misura ed alle Superfici Minime*, 1996 (reprint 2000).
E. VESENTINI, *Introduction to Continuous Semigroups*, 1996 (second edition 2002). ISBN 978-88-7642-258-4
C. PETRONIO, *A Theorem of Eliashberg and Thurston on Foliations and Contact Structures*, 1997. ISBN 978-88-7642-286-7

Quantum cohomology at the Mittag-Leffler Institute, a cura di Paolo Aluffi, 1998. ISBN 978-88-7642-257-7

G. BINI, C. DE CONCINI, M. POLITO, C. PROCESI, *On the Work of Givental Relative to Mirror Symmetry*, 1998. ISBN 978-88-7642-240-9

H. PHAM, *Imperfections de Marchés et Méthodes d'Evaluation et Couverture d'Options*, 1998. ISBN 978-88-7642-291-1

H. CLEMENS, *Introduction to Hodge Theory*, 1998. ISBN 978-88-7642-268-3

Seminari di Geometria Algebrica, 1998-1999, 1999.

A. LUNARDI, *Interpolation Theory*, 1999. ISBN 978-88-7642-296-6

R. SCOGNAMILLO, *Rappresentazioni dei gruppi finiti e loro caratteri*, 1999.

S. RODRIGUEZ, *Symmetry in Physics*, 1999. ISBN 978-88-7642-254-6

F. STROCCHI, *Symmetry Breaking in Classical Systems*, 1999 (2000). ISBN 978-88-7642-262-1

L. AMBROSIO, P. TILLI, *Selected Topics on "Analysis in Metric Spaces"*, 2000. ISBN 978-88-7642-265-2

A. C. G. MENNUCCI, S. K. MITTER, *Probabilità ed Informazione*, 2000.

S. V. BULANOV, *Lectures on Nonlinear Physics*, 2000 (2001). ISBN 978-88-7642-267-6

Lectures on Analysis in Metric Spaces, a cura di Luigi Ambrosio e Francesco Serra Cassano, 2000 (2001). ISBN 978-88-7642-255-3

L. CIOTTI, *Lectures Notes on Stellar Dynamics*, 2000 (2001). ISBN 978-88-7642-266-9

S. RODRIGUEZ, *The Scattering of Light by Matter*, 2001. ISBN 978-88-7642-298-0

G. DA PRATO, *An Introduction to Infinite Dimensional Analysis*, 2001. ISBN 978-88-7642-309-3

S. SUCCI, *An Introduction to Computational Physics: – Part I: Grid Methods*, 2002. ISBN 978-88-7642-263-8

D. BUCUR, G. BUTTAZZO, *Variational Methods in Some Shape Optimization Problems*, 2002. ISBN 978-88-7642-297-3

A. MINGUZZI, M. TOSI, *Introduction to the Theory of Many-Body Systems*, 2002.

S. SUCCI, *An Introduction to Computational Physics: – Part II: Particle Methods*, 2003. ISBN 978-88-7642-264-5

A. MINGUZZI, S. SUCCI, F. TOSCHI, M. TOSI, P. VIGNOLO, *Numerical Methods for Atomic Quantum Gases*, 2004. ISBN 978-88-7642-130-0

The manufacturer's authorised representative in the EU is Springer Nature Customer Service Centre GmbH, Europaplatz 3, 69115 Heidelberg, Germany. If you have any concerns regarding our products, please contact ProductSafety@springernature.com

Printed and bound by CPI Group (UK) Ltd, Croydon, CR0 4YY
23/03/2026
02076446-0011